GLOBAL
TELECOMMUNICATIONS
POLICIES

Recent Titles in
Contributions in Economics and Economic History

Global Telecommunications Policies

The Challenge of Change

EDITED BY
MEHEROO JUSSAWALLA

Contributions in Economics and Economic History, Number 148

Greenwood Press
Westport, Connecticut • London

Library of Congress Cataloging-in-Publication Data

Global telecommunications policies : the challenge of change / edited
by Meheroo Jussawalla.
     p.   cm. — (Contributions in economics and economic history,
ISSN 0084-9235 ; no. 148)
  Includes bibliographical references and index.
  ISBN 0-313-28865-8 (alk. paper)
  1. Telecommunication policy. I. Jussawalla, Meheroo.
II. Series.
HE7645.G58 1993
384'.068—dc20   93-16650

British Library Cataloguing in Publication Data is available.

Library of Congress Catalog Card Number: 93-16650
ISBN: 0-313-28865-8
ISSN: 0084-9235

First published in 1993

Greenwood Press, 88 Post Road West, Westport, CT 06881
An imprint of Greenwood Publishing Group, Inc.

Printed in the United States of America

The paper used in this book complies with the
Permanent Paper Standard issued by the National
Information Standards Organization (Z39.48–1984).

10 9 8 7 6 5 4 3 2 1

Contents

Introduction

Meheroo Jussawalla

A decade after the divestiture of American Telephone & Telegraph (AT&T), many countries are reexamining the wave of deregulation that swept across the countries of the Organization for Economic Cooperation and Development (OECD). Although competition for telecommunications services has been driven by dynamic changes in technology and sophistication of user demand, not all countries are convinced of the benefits of privatization and liberalization, and even if they are, their economies may not be able to support a fully privatized system, particularly in the area of universal basic service. The telecommunications revolution has wrought unprecedented and powerful changes that have been compressed in a short time span and had a phenomenal impact on society. The economic effects of this revolution are beginning to emerge in a significant manner. The industry is facing massive upheavals that challenge established institutional and industrial structures of sectors other than telecommunications.

This rapid growth of electronic highways radically altered the concept of regulators emerging as policy makers. Rule making has become problematic in an environment in which innovations spread across political frontiers even before their application is tested in their country of origin. National and international regulatory regimes have been left behind in the race between technology and industry regulation. The transition from monopoly to competition has been ascribed to factors such as the expansion of business communications, user demand for customized networks, tariff distortions caused by monopoly use of cross subsidies, the trend to bypass the public network, and the growth of electronic data interchange.

From a segregated assembly of national monopolies protected by national governments, Post, Telephone, and Telegraph (PTT) authorities in developed and developing countries were forced to come to terms with a system of changing value-added services that did not carry assurances of guaranteed markets. By the mid-1980s, internationalization was the order of the day and fierce global competition was let loose among vendors who in the past ruled over established fiefdoms of national markets. Indeed, the changes in the industry were so explosive that for the first time, communications were recognized as a strategic underpinning of civilization (McClelland, 1992).

Regulatory economics consists of a number of distinct traditions regarding both positive and normative aspects of the regulatory process: the public interest theory, the capture school, the cooperation theory, and the political economy of deregulation (Snow & Jussawalla, 1989).

Public interest regulation is based on the assumption that a monopolist produces a single output under economies of scale. The monopolist can, in the absence of competition, charge predatory prices by producing smaller quantities of that output to the detriment of the consumers. The goal of regulation in such a case is to require the firm to increase the output, reduce the prices, and recover a fair rate of return on investment, thereby protecting the public interest.

The capture theory emanated from the Chicago school and was based on general flaws in the regulatory mechanism. It suggested that the regulators were "captured" by the industry they were regulating and that regulations were framed to protect interest groups rather than the user of the product or service. Mosco (1988) also subscribes to a similar theory of regulation by experts rather than by political parties and believes that the experts represent special interests, thereby manipulating societal values.

The school that advocates cooperation between the public and private sectors traces its roots to the English cooperative movement of the nineteenth century. The International Telecommunications Satellite Organization (INTELSAT) until recently held a monopoly over international satellite communications. It is a global intergovernmental consortium and is organized financially as a cooperative of owners and users. The French telecommunication authority Directorate General de Telecommunications (DGT) has stressed cooperation as an alternative to competition in obtaining the goals of economic efficiency and social welfare.

The political economy of deregulation school is better known for its recent application to information technology (IT) in industrially advanced countries. It stresses the role of the state together with interest groups and of the general international environment in affecting the outcome of policy debates.

Central to the theory of regulation is the concept of a de facto or natural monopoly that enjoys economies of scale and scope because of the multiproduct nature of IT. Seminal works by Baumol (1982) and Baumol, Panzar, and Willig (1982) have emphasized the definition of natural monopoly as a property of the cost function. With economies of scale, average cost decreases as the level of output increases. When a single firm is able to manufacture a combination of outputs, economies of scope obtain that are superior to a number of firms producing the same outputs individually. A natural monopoly is said to be sustainable if it can frustrate any attempt at entry into its product market by a smaller or more specialized rival. However, if the market is contestable and entry to it is not legally barred, such contestability serves as a constraint on the monopolist and makes the operations more responsive to user needs. Regulation then would become unnecessary.

In the United States, the successive shifts and reverses in Federal Communications Commission (FCC) policy since 1982 show that the policy makers have found it difficult to strike a proper balance between allowing smaller competitors to flourish and not placing too many restrictions on the dominant supplier. The industrial structure on the supply side has to adjust to the loss of captive markets due to deregulation, increasing costs of research and development (R&D) for new products, and rapid changes in innovation cycles.

This volume covers issues such as the possibility of separating policy making from regulation within a framework of politically accountable institutions when countries vary in their customs, laws, and practices of ownership. For example, in the Asia Pacific countries, telecommunications have been ascribed a low priority in development investment and have been traditionally administered through ministries of transportation and communication. Even in Japan until 1985 there was considerable government intervention in the operations of the Nippon Telephone and Telegraph Company (NTT) by the Ministry of Post and Telecommunications (MPT). Such policies were based on the public interest theory of regulation, which stipulated that the intervention was in the interest of protecting those whose voice would not otherwise be heard (Hills, 1991). It was a trade-off between predatory pricing and protection of the monopoly. In the United States, a regulation by the FCC of AT&T was justified on the basis of universal access. Stigler (1971) disagreed with this view inasmuch as the monopoly was favored rather than the users of the service, which was the view of the capture theorists.

The group of national monopolies in countries around the world, PTTs, have come to terms with a free market environment in which their market share is not guaranteed. For the first time, countries have come to realize that an inadequate communications infrastructure will prevent the participation of their national economies in global markets. Export-oriented countries like the newly industrialized economies (NIEs) have come to realize this aspect of the information revolution and have tuned their policies to keep their leading sector competitive. The cause of such decision making is that the IT industry together with broadcasting has become a multitrillion-dollar industry worldwide.

This volume discusses the aftermath of such milestone events as the breakup of AT&T, the duopoly in the United Kingdom, the privatization of NTT, and their impact on national and international markets. The frequency spectrum debates, the standardization problems, and the entry of telephone companies into the cable television market were not foreseen in the mid-1980s. Even the Thatcher government perhaps was not fully aware of the revolution it triggered in telecommunications, nor were the Japanese fully cognizant of the effects of floating NTT stocks in 1985. The revenue streams from privatization soon became apparent. It was also an opportunity to modernize antiquated infrastructures. The variety of liberalization and privatization policies have been covered by the chapter authors in this book, who in their own rights either are distinguished architects of policies or have great knowledge on these issues.

The Pacific and Asian countries such as New Zealand and Malaysia have sold their PTTs or privatized them to raise cash to either fund external debts or expand their networks. In Australia the satellite-based and publicly owned carrier Aussat has been sold to form a new company (Optus) with foreign collaboration in ownership. In some OECD countries, such as France and Germany, regulation has been separated from the official policy-making structure and the PTT is being operated on commercial lines. However, it is ironic that the opening of markets has been accompanied, in some instances, by increased regulation (such as of AT&T, British Telecom, and NTT). This regulation in the interest of social rather than economic goals is widely discussed by the contributors in this volume.

The United States model of regulation by experts, with emphasis on separating regulation from policy making and the transparency of rules enacted, is the preferred model in Europe (Hills, 1991). Furthermore, the U.S. model is one of many layers of private operators interconnected and regulated at the local, state, and national levels. In Europe there are not so many layers of control requiring integration. For example, in France the objective of the government monopoly is to provide an integrated network throughout the country; France boasts of one of the world's highest rates of network digitization, the largest packet switching network, and national availability of Integrated Services Digital Network (ISDN). Many of these services have been provided in anticipation of user demand (Scherer, 1992). The same has been true of Singapore, which introduced videotex initially to stimulate user demand (Jussawalla, 1992).

Within the European Economic Community (EEC), an IT task force was formed within the bureaucracy to develop in all the member countries a common telecommunications policy that would cover computers, data bases, electronics, and network integration. The market orientation for strengthening the industry was the objective of the Directorate General (DG) XIII. The idea of the Single Market 1992 further inspired the desire for a unified policy for equipment and services. The announcement of moving toward competitive structures was first made by the Green Paper in June 1987. The objective of the European Commission was that fragmented markets as they existed should disappear, and this led to a new balance of power between the DG IV (in charge of competition) and the DG XIII (in charge of telecommunications). The issue before the Commission is whether the telecommunications industry should be governed by EEC rules of competition or it should create a level playing field and allow the market to take care of competition. The Commission has followed the example of the Ministry of International Trade and Industry (MITI) in Japan by promoting successful R&D programs in the 1980s, such as RACE (Research in Advanced Communications Technologies in Europe) and ESPIRIT (European Strategic Program for Research in Information Technology). The concept of regulations posed no problems in Europe, but the directives of the EEC Treaty have been subject to legal controversy in the telecommunications field. Article 100 of the EEC Treaty is crucial to the speedy implementation of the harmonization and coordination required to achieve the goals enunciated by the Green Paper (Scherer, 1990). The concept of the Open Network Provision was specified in the Green Paper in order to harmonize the terms and conditions of access network facilities for nonreserved services (for example, basic service). The reforms in Europe also aimed at liberalizing satellite-based services.

Thus, it can be seen that the European policy is different from that of Japan and the United States to the extent that network infrastructure is a reserved service and facilities-based competition is permitted for value-added services. The consensus within the European Commission's policies is currently being eroded by new developments in the satellite-based networking services and the urgent need to establish connections between Eastern and Western Europe, the former being handicapped by archaic infrastructure.

The above theoretical and comparative background is provided to mould together the chapters in this volume and give an overarching view of the changing trends that challenge telecommunications policy makers. For example, compared with Europe and Japan, telecommunications reforms in Latin America started with a narrow focus on privatization of PTTs, but soon the governments realized that a more-comprehensive approach to restructuring was needed. Reforms in Latin America have been successful in mobilizing foreign capital, obtaining the experience of successful operators in the industry, and opening access to the latest technologies. The reforms have followed a pattern similar to that in most other countries by selling a controlling interest in the state enterprise to a private or foreign owner and the balance to a number of domestic investors, reducing the scope and limiting the duration of the monopoly held by the former PTT, opening some segments of the telecommunications business to international competition, and starting new regulatory supervision. The state PTTs in Argentina and Mexico were privatized with record speed, faster than those in the United Kingdom and Japan. ENTEL (the telephone company of Argentina) and TELMEX (the telephone company of Mexico) became the models for other industry restructuring in Argentina and Mexico, respectively.

Likewise, in Southeast Asia, including Malaysia, Korea, Taiwan, and Singapore, deep structural reforms are in progress, but we cannot conclude from this that all countries that are privatizing were not performing well before. For example, Singapore had the most sophisticated networks offered by its statutory organization, Singapore Telecoms. Thailand is beset with poor quality domestic and international services. Gradually Thailand is permitting its data- and satellite-based services to be owned and operated by private companies, such as Hitachi, which provides cellular mobile services in Bangkok. The fact remains that the opening of telecommunications markets to competition is a matter of time as pressures both economic and political are being generated alongside user demand for integration with global networks that support international trade.

This collection aspires to offer a wide scope of coverage of telecommunications reforms over the 1980s and through the first half of the 1990s. The aim is to provide the benefits of different views of the recent national and international developments in telecommunications policies and to provide research materials in a single volume. The authors deal with current regulatory challenges in a timely and unique manner. This book holds for the reader a wealth of information rich in comparative policy lessons and approaches.

We have become aware that these regulatory changes have become more or less unending. The U.S. Congress in September 1992 passed a bill to regulate cable networks for overcharging the users. President Bush vetoed the bill, and Congress overruled the veto. Similarly, problems have arisen in the United States with international telephone calls. Discount call companies are sprouting globally to provide lower rates for international calls than rates provided by the recognized common carriers. The world market for international calls was worth $50 billion in 1991. The U.S. market for international calls is open to competition; that of other countries is not. Calls out of the United States are far cheaper than calls into it. The discount companies exploit this difference by routing calls from foreign subscribers to computerized switches in the United States, undercutting normal rates by as much as one-third. International Discount Telecommunications (IDT), an example of such service providers, sells discounts to the World Bank's African offices (*The Economist*, September 12, 1992).

The changes taking place in the United States after the divestiture of AT&T are described in the first chapter of the book. Henry Geller, who headed the U.S. government's National Telecommunications and Information Administration, has succinctly described the role of the FCC in meeting the insatiable demand for data transmission requiring digital technology. With broadband integrated services digital networks in the offing, many new services will appear, such as links between office computers, rapid access to libraries of information, video phones in homes and offices, and the smooth working of information highways. The regulatory means of achieving policy goals and of imparting efficiency to these superhighways are dealt with clearly in this chapter. New issues of spectrum allocation and high-definition television (HDTV) standards are discussed. The impetus given by the FCC in coordinating standards for HDTV has resulted in a technology that promises to overtake the Japanese one, which uses analogue (as contrasted to digital) systems.

France has moved from a state monopoly to liberalization very gradually, as described in the evolution and trends of French policy by Jean-Pierre Chamoux. The author served on the regulatory law-making body of the French government and edits a journal in the field called *Le Communicateur*. After tracing the historical background of the French telecommunications industry, the author examines carefully the drive toward European integration, the creation of a separate regulatory body, and the rebalancing of tariffs. These changes are examined in the context of other European countries' policies, which appear far more rapid than those in France. The Minitel system has gained recognition in Europe and spurred videotex systems worldwide. The future course of events is tied in with the formation of the Single Market and the stakeholder interest that France Telecom has in its operations. The author is not optimistic about privatization of the France Telecoms in the near future.

Deutsche Telekom's policy is analyzed in the context of the larger picture of the Single Market and the various bodies constituted to make it succeed. The Witte Commission's Report first initiated the movement toward liberalizing the Bundespost. The chapter is authored by Klaus Grewlich, who is the Director General for International Relations of the Deutsche Telekom. After a scholarly and empirical discussion of harmonization and liberalization policies in the context of the Green Paper, the author deals with the challenges faced by the European community in general and Germany in particular, along with the issue of integration of networks with the eastern part of Germany, which is saddled with a less-efficient infrastructure.

Recent developments in telecommunications policy in the United Kingdom are comprehensively explained in the chapter by Adrian Norman, who used to be a consultant with Arthur D. Little and now owns his own consulting practice. His analysis elaborates on the difficulties encountered in making institutional adjustments to keep pace with changes in the telecommunications sector. According to the author, technologies have converged rapidly, but lack of parliamentary understanding of such changes has slowed the pace of policy changes to adjust to technological ones.

Japan has been investing heavily in digital highways, with adequate support from governmental institutions. Hajime Oniki describes the changes in Japan since the reform of 1985. Specialized in the economics of telecommunications, Oniki provides a detailed analysis of the major impact of the Business Law on the industry over the past seven years. His findings are substantiated by charts showing the effects on telephone tariffs and the pricing of other services. This chapter provides important insights for vendors from other countries regarding entry to the Japanese market. (After the writing of this chapter, NTT put one more tranch of its stock on the market in a move toward greater privatization.)

The regulatory process in Canada is historically charted by William Melody and Peter Anderson. They show how Canadian policy has lagged behind that of the United States and how this has affected the competitiveness of the telecommunications industry in Canada. The position now may change with the challenge from the free trade agreement with the United States and Mexico (North American Free Trade Agreement [NAFTA]).

Although Canada has been slow in adjusting its policies to innovative IT, Australia has taken a giant leap forward and privatized its PTT. The Australian developments are accurately described by Donald Lamberton. In a penetrating analysis, the author examines the information intensity of Australian society and recounts the events that led to the current rapid changes in policy. He is critical of the lack of research in the selection of new technologies.

The complex policy structures in Latin America are dealt with extensively by Raimundo Beca. He begins with the asymmetric processes of privatization in Chile, Argentina, and Mexico and then examines the cases of Brazil and Venezuela. The vast wealth of information in this chapter is substantiated with charts of comparative performances.

We find that these countries are forging ahead in their technology policies when compared with the slow progress in Africa, as described in the chapter by Raymond Akwule. With fragile economies and social dualism prevalent in many African countries, the PTTs have not been able to adequately cope with providing new technologies. Even basic telephone networks are not available to a vast majority of people on that continent.

By contrast, the People's Republic of China, under a highly centralized system and with a vast hinterland populated by the world's largest concentration of people, appears to be managing its policies better than many developing countries. Lin Sun shows how demand is far outstripping supply for telephones in China and how the MPT is handicapped by an acute shortage of funds for boosting supply. Even so, China has provided fiberoptic cable for its public switched networks and satellite communications for domestic and international users. Perhaps its greatest achievement has been its launch vehicle for satellites, which is now widely being used by Asian and European countries. Called the Long March rocket, it has successfully launched many communication satellites, the chief of which being its own Asiasat satellite, in which China has one-third ownership. It is a system that is providing television programs to the developing countries of South and Southeast Asia.

The regulatory policies of the Association of Southeast Asian Nations (ASEAN) are analyzed in the context of the move toward regionalism that has overtaken the developed countries of the world. Although there exist different levels of economic and social development and although political systems diverge, there is considerable cooperation that characterizes decisions regarding telecommunications. In the ASEAN, as in the Asian NIEs, privatization and liberalization of telephone networks are forging ahead as new technologies are introduced to tie these economies to the global markets. They now constitute one of the world's fastest growing markets for telecommunications equipment and services. They are also export-oriented countries that are concerned about the new free trade blocs like the European Single Market and NAFTA.

Heather Hudson brings her insights to the chapter on development communications. She has shown how policies need to be restructured to provide basic and thin route communications to remote areas.

The volume is brought to a close by the second of Grewlich's chapters, which lays out the agenda for the 1990s for cooperative communications policies, without which the benefits of the information revolution cannot be received equitably by the countries around our shrinking globe. For the "global village" of Mac Luhan to become a reality, people in the remote and rural areas must have at least basic telephone service. According to the Secretary General of the International Telecommunication Union, Pekka Tarjanne, the right of access to telecommunications is a fundamental human right.

The question for the 1990s is whether the industry can create a globally competitive telecommunications infrastructure without a greater role for the government. Much depends on the answer. Today's telephone lines have to handle not just voice messages but also torrents of computer data, images, and video. The service economy has an insatiable appetite for communication capacity such as videoconferencing and distribution of information-based services. The technological challenges of blanketing nations with information superhighways are daunting, but equally challenging are the regulatory regimes that keep the information flowing. This volume addresses these challenges from national and international points of view. Regulators and policy makers continue to be involved in a race between technology and policy in order to best safeguard the interests of the corporate and individual users and to provide affordable services to the low-income countries.

REFERENCES

Baumol, W. J. (1982). Contestable markets: An uprising in the theory of industry structure. *American Economic Review*, 72(1), 1–15.

Baumol, W., Panzar, J., & Willig, R. (1982). *Contestable markets and the theory of industrial structure*. New York: Harcourt Brace Jovanovich.

Discounted telephone services. (1992). *The Economist*, September 12.

Hills, J. (1991). Regulation, politics and telecommunications. In J. P. Chamoux (Ed.), *Deregulating regulators* (pp. 100–01). Amsterdam: IOS Press.

Jussawalla, M. (1992). Is the link still missing? *Telecommunications Policy*, August/September, 485–503.

McClelland, S. (1992). The international dimension: PTTs. *Telecommunications*, June, 31–37.

Mosco, V. (1988). Towards a theory of state and telecommunications policy. *Journal of Communication, 38*, 107–24.

Scherer, J. (1990). Regulatory instruments and EEC powers to regulate telecommunications services in Europe. In D. Elixman & K.-H. Neuman (Eds.), *Communications policy in Europe* (pp. 236–48). Berlin: Springer Verlag.

Scherer, P. (1992). Perspectives on world telecom reform. *TDR*, May/June, 28–29.

Snow, M., & Jussawalla, M. (1989). Deregulatory trends in OECD countries. In M. Jussawalla, T. Okuma, & T. Araki (Eds.), *Information Technology and Global Interdependence* (pp. 21–35). Westport, CT: Greenwood Press.

Stigler, G. (1971). The theory of economic regulation. *Bell Journal of Economics and Management Science*, Spring, 3–21.

1

Changing Technologies and the Role of the FCC in the United States

Henry Geller

This chapter sets forth highlights of the role of the Federal Communications Commission (FCC) in effecting structural or regulatory adjustments to deal with rapidly changing technology in the U.S. telecommunications sector. The goals of FCC policy are twofold: to have telecommunications make a maximum contribution (1) to efficiencies or improved productivity, so essential to all industries in this era of global competition; and (2) to the quality of life for U.S. citizens in such areas as universal access, education, health, and an informed electorate.

PRINCIPAL REGULATORY MEANS FOR GOAL ACHIEVEMENT

Reliance on Competition

Competition is the norm in the United States because it drives prices to marginal cost levels and spurs efficiencies and innovation, including those by the smaller entrepreneur filling niches. The change from copper wire and electromechanical switches to microwave, satellite, coaxial, and photonic transmission and, above all, to the computer (integrated circuitry) for both transmission and switching now makes competition feasible throughout the telecommunications industry.

Thus, the trend is now to digital technology as the clear base of information networking. Information is now created, stored, manipulated, and transmitted in digital form; this includes all computers and imaging systems and all modern transmission and switching systems in telecommunications networks. The cost of this digital equipment continues to fall substantially, and the power of the digital processing continues to increase — almost doubling every year — so that by the end of the century, it will be a thousand times more powerful.

Photonic transmission capacity is also doubling every year, with the process continuing until the year 2000. Such transport, comfortably into the billion-per-second range, continues to stimulate higher and higher speed transmission of information, with switching technology closing in at trillion-per-second packet-switching systems.

This technology base has led to a convergence of previously separate markets — telecommunications, computers, and television. The multimedia computer, now coming on stream with the more powerful processing, will embrace them all — voice, data, graphics (imaging), and video, and this new computer, to be fully effective, needs correspondingly powerful telecommunications links.

Avoidance of Regulated Competition

Regulated competition should be avoided when effective competition is achieved. When the policy of open entry results in effective competition, it is important to deregulate that sector. To continue to regulate in such circumstances is the worst of all possible worlds and frustrates full achievement of the benefits of the competitive environment. However, as one U.S. senator stated, "all each industry seeks is a fair advantage over its rivals," and that can translate into continuing pressure for regulation in order to handicap the rival.

In accordance with this policy, the FCC has completely deregulated the Customer Premises Equipment (CPE) field, so that now anyone can attach certified equipment to the telephone system, just as in the case of the power line. This has been an outstanding success. The FCC has moved to deregulate the enhanced service field, that is, a carrier service that involves data processing services such as call storage and forwarding and protocol and code conversion. This makes sense because the field is one of effective competition, both from CPE (which can be the functional equivalent) and from noncarriers. However, unlike in the case of CPE, the court has ruled that the FCC lacks the power to preempt intrastate regulation of such enhanced services in California v. FCC, 905 F.2d 12117 (9th Cir. 1990). This illustrates a flaw in the U.S. policy-making process — the absence of a fully effective federal "captain."

The FCC has also moved to largely deregulate the interstate toll field by reducing greatly the regulatory burden on the one interstate toll carrier still under such scrutiny, American Telephone & Telegraph (AT&T). This has met some resistance in the industry, reflected then in Congressional concern. Significantly, half the states have wholly deregulated the intrastate toll area (including AT&T), with considerable competitive benefits and no adverse consequences.

Incentive Regulation

Incentive regulation should be used rather than rate of return (ROR). Even with strongly competitive policies, the local exchange area remains one where the local exchange carrier (LEC) still has monopoly power, because it is the ubiquitous carrier reaching all businesses and residences. There is burgeoning competition, especially in providing access for large businesses to the interexchange toll carriers or in fiberoptic rings in the downtown business areas, but the overall system, including the ubiquitous subscriber loop, gives the LEC monopoly power. There is thus the need to protect the monopoly ratepayer.

The FCC seeks to meet that responsibility in a way that promotes the goal of spurring efficiencies and innovation. It has, therefore, turned to a price cap regulatory regime for interstate operations (for example, AT&T, the divested Bell Operating Companies [BOCs]) similar to that undertaken in the United Kingdom, that is, appropriate baskets of basic service are subject to price caps that are adjusted over a four-year period for inflation and productivity indexes. This avoids the cost-plus nature of ROR regulation that can inhibit efficiency gains (that is, the carrier gains nothing from introducing efficiencies if it is close to its authorized ROR). This price cap regime is not deregulation; rather, it institutionalizes regulatory lag, in effect saying to AT&T or the LEC, "You have four years in which if you are very efficient, you retain the earnings thus gained; at the end of that period, there is an adjustment in which the ratepayers then gain from the efficiencies — hence a win-win situation because of the incentive to be efficient."

Avoidance of Market Segmentation and Competition Suppression

Market segmentation or the suppression of competition should be avoided. The present chairman of the FCC, Alfred Sikes, has stressed the importance of avoiding suppression of competition; he strongly believes that each entity should be allowed to use its facilities in a way that affords a maximum contribution to telecommunications competition. Thus, a cable television system should be allowed to use its infrastructure to provide other telecommunications services such as local access to carriers, local transport of data or voice, or personal communications services. Similarly, the broadcaster should be permitted to use subcarriers to engage in telecommunications operations such as paging. The local regulators (public utility commissions [PUCs]) often resist this because of their desire to maintain local monopolies in order to promote social purposes, a matter dealt with in the discussion of the subsidy scheme.

The antitrust suit resulting in the divestiture of AT&T imposed serious market segmentation restrictions on the divested BOCs. They were barred from providing information services (the enhanced or data processing–type services referred to above), manufacturing telecommunications equipment, or engaging in interexchange operations between defined areas called local access and transport areas (LATAs). The FCC believes strongly that these restrictions deny consumers significant competitive benefits from roughly one-half of the telecommunications sector in the United States and should be abolished.

The FCC position is now supported by the Commerce and Justice Departments, and the restrictions on information services and manufacturing are being reviewed, the former by the antitrust court and the latter in pending legislation in Congress. It appears likely that by the mid-1990s these two restrictions will be gone and that by the end of the decade, the final restraint (on intra-LATA interexchange operation) will also end. However, once again, the conflict between the positions of the antitrust court and the FCC points up the absence of the federal "captain" and the failure of Congress to act to resolve the matter.

Insurance of Equal Competition

A level playing field should be insured for the competitors of the BOCs. The BOCs are now permitted entry into some information services that do not involve content generation or content manipulation (for example, protocol and code conversion, voice storage and forwarding, voice mail). Because they remain the monopoly network providers, it is necessary to regulate the interface between the monopoly and the competitive information services for two reasons: to insure against improper cross-subsidization (that is, improperly assigning some joint and common costs to the monopoly side that really should be allocated to the competitive enterprise) and to insure that the competing information service providers (ISPs) are afforded fair interconnection to the parent monopoly network.

The use of price caps is certainly a help in insuring against improper cross-subsidization, although not a complete solution, because the cap is reviewed at four- or five-year intervals (as stated, it really institutionalizes regulatory lag). The FCC in its 1980 Computer Inquiry II decision required a separate subsidy for the competitive enterprise, thus very largely eliminating any problem of improper cross-subsidization, because there would be no joint and common facilities. In its Computer Inquiry III decision, the FCC abandoned this structural approach because of claimed inefficiencies and in its place instituted a detailed accounting scheme. The 1990 California case referred to above found that the FCC had erred in this latter decision and remanded the issue for reconsideration by the agency, which has issued a notice of proposed rule making. The issues of accounting versus structural approaches and of the economies of scope from integration remain a difficult and unsettled matter. Further, here again, it is not clear if the FCC can impose its decision upon the states as to purely intrastate enhanced service operations (assuming such wholly intrastate operations can exist as a practical matter).

As to fair interconnection for the ISPs, the FCC in the Computer Inquiry II decision required that the BOCs afford comparably efficient interconnection (CEI) to ISPs for any enhanced service they wish to enter. Thus, CEI requires that a specific interface be designed for each enhanced service such as voice storage and forwarding or voice mail. This is time consuming and can delay provision of new services; further, to obtain interconnection, an ISP must disclose its proposed operation to the BOC, a potential or actual competitor. The FCC thus in the same report called for a generic approach — open network architecture (ONA).

ONA represents an unbundling of the network into its various elements or nodes; the user, such as an ISP, could then select the element or node that it wanted (for example, transport, switching, billing, use of the intelligent network [SS7] as automatic number identification [ANI] or locator function). The FCC called for the submission of ONA plans delineating basic service elements. The BOCs responded with plans that entail basic service arrangements involving the above elements but wedded to some transport facet. The real unbundling of the network has not yet been accomplished. The Commerce Department has urged that there be a separation of switching, the subscriber loop transport, and transport between central offices.

Some states, such as New York, California, and Illinois, have acted in this area to afford competitors operating fiberoptic rings or systems of businesses either colocation or virtual colocation of their transport facilities in the BOC central office. The FCC, following this lead, has proposed the same kind of interconnection for interstate facilities. These interconnection or ONA issues are of the greatest importance and will undoubtedly receive intensive consideration in the 1990s. With the proliferation of networks and service providers, interconnection rules must be a primary focus of the FCC.

Rationalizing the Subsidy Scheme

In light of the increasing competition and its associated matters such as true ONA, there is a corresponding need to rationalize the subsidy scheme. Under the AT&T monopoly arrangement, large subsidy elements were introduced — for example, arbitrarily shifting costs to the interstate toll sector to reduce costs in providing local service, thus contributing to universal service. However, competitors are naturally attracted to segments in which prices are artificially high because of the social purpose "tax." Stated differently, once competition is let loose in the system, large subsidy schemes become infeasible.

This is not to say that subsidies are not needed. They clearly are for the poor or those in very high cost rural areas, but the old system of subsidizing everybody, including the rich or middle class, must go. The subsidies should be targeted to those who need them, just as in the case of food or energy. The FCC provided leadership here by largely eliminating the arbitrary "tax" on interstate calls (substituting a subscriber line charge), making the interstate subsidies explicit, and targeting them to the poor or high cost areas. There is the need for the states to follow suit and rationalize prices on the local level.

PENDING ISSUES

The foregoing are the main principles that have been guiding FCC action to deal with the dynamic nature of the telecommunications field. It may be most useful to illustrate their applicability by focusing on two important and pending issues, one in the wire field and the other wireless. The first issue involved policy matters regarding the provision of a universal switched broadband (fiberoptic) network. The second is the provision of personal communications service (PCS), the use of digital radio for tetherless telecommunications services. Indeed, the two issues are related, because over the next generation, there may well be a switch, with television moving to the wire (fiber), and telephone (voice and data) moving to radio.

Provision of a Universal Switched Broadband Network

The United States today has a broadband infrastructure, cable television, that passes 86 percent of all television households and is subscribed to by 60 percent of such homes. Further, cable is on the point of strong capacity expansion. The industry intends to rebuild its systems using fiberoptic "tree" architecture with coaxial "branches" or drops into the home continuing; this will give it the capacity of

150–200 channels, and with digital compression techniques that are now in the offing, it could reach the 500-channel mark.

However, as recent FCC proceedings and Congressional hearings have made clear, cable, although it can and should make a very substantial contribution to the U.S telecommunications infrastructure, has several drawbacks. First, it is not a switched system; rather, it employs "tree and branch" architecture. Second, it is not a common carrier. It is a downstream video packager that claims the right to determine what services will be allowed to reach the subscriber. Its decisions in this respect are based on self-interest — maximizing its profit and, in light of significant vertical integration into programming, protecting its associated programming channels.[1]

Thus, there is only one 24-hour cable news channel, CNN, not because the market so decrees (NBC sought to provide a competing news channel) but because the large cable operators like TCI and Time-Warner, who own a substantial stake in CNN, do not want any competitors to CNN. This may be good business, but from a First Amendment point of view, it represents a horror case, because U.S. democracy is founded on the First Amendment and its underlying premise that the people of the United States should receive information from as diverse and antagonistic sources as possible.

Finally, there is great concern because of the monopoly position cable now has for the distribution of its large cluster of services (for example, pay channels, advertiser-based cable programmers, and hybrid programmers [part advertiser, part subscriber fee–based]). Questions have been raised as to rate "rip-offs" and the quality of technical or consumer service, stemming from this monopoly situation.

The FCC, the Commerce Department, and some in Congress have looked to the telephone companies for a solution to the above problems. The telephone company is vitally interested in fiberoptic transmission. Fiber has replaced copper in trunking, is being extensively used in feeder plants, and is not "proving in" to replace copper in the subscriber loop, first to the curb or pedestal (estimated to begin in 1992–93) and then to the home (estimated for 1995). It will be used for new developments and to replace worn-out copper. However, this replacement takes place at a rate of roughly 3 percent per year, thus entailing a 30-year or longer time period. The critical issue for policy makers in Congress and at the FCC and state PUCs is whether to authorize accelerated deployment of fiber, so that the task will be completed within one generation or earlier (for example, about 2015).

In the author's view, such accelerated deployment is warranted for two reasons. First, it will attain much earlier the vitally needed First Amendment infrastructure allowing video publishing over a common carrier. Second, it will also meet the coming need for local switched broadband networks. As noted, the powerful multimedia computers, combining data, voice, imaging, and video, are coming on stream and, to make their full contribution to economic efficiencies and quality of life, need a switched broadband highway into every research institute, business, and home. The federal government is now moving to support linking all research institutions. Many businesses have developed their own fiber networks but still need the public switched network to reach suppliers or customers, who are located throughout the city and its environs. It also is argued that the home needs this network because of new services in areas like education, health, and home work stations.

In light of past U.S. experience with cable and videotext, there is no way now to forecast the home demand for services involving switched broadband, but it does seem prudent, in view of the great potential and the above business needs, to plan the local broadband system so that it is readily and gradually upgradeable to a switched status (that is, through the addition of photonic or other improved switches and optoelectronic equipment at the appropriate ends). This consideration should be taken into account in planning the initial system and in determining the accelerated deployment schedule.

Along with the issue of accelerated deployment, Congress and the FCC have ongoing proceedings looking at permitting the LECs to offer cable television services through a subsidiary. Such offerings are now prohibited by law except for LECs operating in sparsely populated areas.

Provision of Personal Communications Service

Just as is the case in Europe and Japan, the United States (FCC) has a proceeding underway to introduce PCS — communications services based on digital radio access technologies that free the individual from the wireline tether, that is, the individual will be able to make or receive calls independent of his or her physical location. Basic to this concept is the use of small, lightweight (seven ounces), low-power portable digital radio handsets (or data terminals) that are inexpensive to purchase ($100 to $150) and to use (for example, $30 per month). Most importantly, just as any phone works anywhere in the network today, the handset should be universal; standards thus are a crucial consideration.

The current cellular mobile radio structure in the United States has the capability to provide PCS to the pubic and will undoubtedly be used for these purposes. However, the cellular system is one of duopoly, limited to 50 megahertz; it is devoted largely to serving vehicular traffic, thus using higher power and much larger cells than would be the case in PCS. It is by no means clear that the great demand projected for PCS can be met through the current cellular scheme. The FCC is therefore proposing to afford much more spectrum for PCS and to open the area far beyond the present duopoly situation in the cellular system.

There were voluminous filings in response to the FCC's 1990 Notice, and the Commission was in the process of formulating a notice of proposed rule making to be issued in late 1992. This second notice will treat the difficult issues of what spectrum will be used and who will be authorized to apply and on what terms. It is therefore not possible to make definitive assertions at this time. What follows is the author's judgment of the probable FCC course of action.

It appears likely that the Commission will employ an open entry policy, just as it did in its action in the air-to-ground proceeding — that is, any entity may apply but must show that it is financially and technically qualified to construct and operate the proposed system. The Commission has indicated that it wants competition both to the cellular duopoly and, even more important, to the subscriber loop, now a monopoly. Indeed, without the advent of strong digital radio competition to the loop, there would appear to be little prospect of eventual deregulation of the local telecommunications sector. There-fore, although such strong competition remains speculative and far off, the FCC policy should be to act now to do all it can to provide that effective competition in the first part of the twenty-first century.

For this reason and because the demand for PCS is thought to be very great, the FCC will look to allocate a substantial amount of spectrum. This poses a difficult issue for the Commission. It is possible to find spectrums higher than five megahertz, but this entails new technological developments, for example, using gallium arsenide. The FCC appears to favor use of spectrums below three megahertz, just as in the case of Europe (and that range was considered at the 1992 World Administrative Radio Conference (WARC).[2] With a few small exceptions, all spectrums below three megahertz have been allocated for use in the United States. Reallocation is a long, hard-fought process that could delay this important service. Congress is now moving to reallocate 200 megahertz from the government service to nongovernment, but again, a considerable time period, on the order of seven years, could well transpire before authorizations on such spectrums could be effected.

It therefore appears likely that the Commission will proceed to use spectrums in the 1.7–2.3 range that are already allocated to, say, fixed microwave and simply make PCS coprimary with the present "grandfatherly" authorizations. The new PCS operators could then work around these existing operations, using time division multiple access (TDMA) or spread spectrum techniques (code division multiple access [CDMA]). If it were found infeasible to engineer the system in this fashion, the FCC would then leave the matter to the marketplace — namely, the PCS enterprise would buy out the existing fixed microwave operators and enable them to move higher in the spectrum. This would result in voluntary reallocation of the spectrum targeted for such flexibility. That might be 1,850 megahertz to 1,990 megahertz or 2,110 megahertz to 2,220 megahertz or, indeed, as proposed by AT&T, the entire range of 1,710 megahertz to 2,290 megahertz, leaving it to the market to decide how much is eventually used for PCS.

Despite this reliance on the marketplace, the Commission would still have most important tasks: to set the interference parameters; to insure the common air interface; to assure interconnection to the LEC network, because many of the PCS operators may not be stand-alone systems (for example, a cable operator connecting customers via PCS to long-distance carriers) but rather must access the local network; and several similar undertakings. There is the complex decision to be made between TDMA or CDMA; it is to be hoped that current experiments will establish whether CDMA is effective in a large-scale urban setting. Eventually this decision must be made by the FCC, which will rely heavily upon industry standards–setting groups; the agency may well have to spur these groups to reach consensus agreement (that will be in all their interests).

Under this scheme, there will be no spectrum set aside for the LECs, as was the case in cellular radio, but the LECs will be permitted to enter the PCS field, because it is just another technology to carry out the same functions — transport and access. Clearly, a condition of such entry will be the provision of fair and effective interconnection of the LEC network to all PCS operators.

The development of PCS under the above scenario will be messy. The earliest entrants will be the cellular operators (who may encounter difficulties overlaying a low-power, microcell PCS system on the high-power vehicular one that must be served) and those who choose to enter via Part 15, utilizing frequencies in which anything goes but there is no protection against interference. However, there are simply not enough spectrums for these early entrants to meet the large vision of PCS that I believe the FCC holds. It follows that there will be a gradual shift to the PCS operations in the spectrum area where a large amount has been set aside.

Further, there are other messy issues to be dealt with. There is the matter of authorization of PCS on a private radio or a common carrier basis, with the line between the two types of services blurring rapidly. If authorized as private radio, the state regulatory commissions have no role, but these commissions might take a great interest in the intrastate aspects of the common carrier PCS operation. Indeed, under proposals of the Bell Regional Companies (Bellcore) and the New York Network Exchange (Nynex), it is possible that digital radio could be used to replace the present copper loop simply because it is more efficient and cost-effective. Such a development would obviously impact local regulation.

These two areas — the universal switched fiberoptic system and PCS — have been discussed simply to illustrate the nature of the regulatory problems confronting policy makers like the FCC in the present situation of fast-changing technology and market conditions. It is essential that policy, the third leg of the stool, keep pace with the dynamic technology and market if the United States is to flourish in this information era of global competition.

NOTES

1.   In September 1992, the U.S. Congress passed a bill to regulate the cable industry in view of overpricing of services. The president vetoed the bill. Congress overrode the veto.

2.   Satellite sound broadcasting received a worldwide allocation of 40 megahertz to 1.45 gigahertz at WARC 1992. The United States and India will use 2.3 gigahertz for their satellite sound systems. The 25-megahertz allocation will take effect in October 1993. For sound direct broadcasts via satellite, the United States chose the S band.

2

French Telecommunication Policy: Evolution and Trends

Jean-Pierre Chamoux

Post, Telephone, and Telegraph (PTT) operations are undergoing significant changes in Europe. The opening of traditional borders, the enlargement of the Common Market, the new pattern of telecommunications services, all influence this evolution. However, national governments do maintain their control on most issues, and each country has finally a proper way to state its policy lines and implement its government views on this sector.

For many foreign observers from the United States, from the Far East, or even from Africa, there is a tendency to assimilate the European policy with a free marketeer approach and with a growing privatization of traditionally state-owned enterprises. The British example is very well-known worldwide and considered as a case study for both privatizing as well as liberalizing the telecommunications networks operations. However, this case is not necessarily the model other countries follow in Europe.

The West German government, for instance, has enacted new rules of law in order to separate the operation of the Post from Telecom in 1989 and to liberalize the service sector around their national public network. However, there is no trace of privatization in that move: the Federal Republic remains the sole owner of the telephone infrastructure except for a duopoly that has been established on cellular mobile telephones for the next digital standard (called Groupe Special Mobile [GSM]). Similarly, a country like Holland has created a public-holding PTT company from January 1, 1989, with subsidiaries to operate and maintain the postal and telephone networks; so far, no shares of these companies have been sold to the public. At the edge of the Common Market, Sweden has had for a very long time a public telecommunication enterprise, Televerket, whose operations and services are being more and more driven by market forces, although no privatization of this public agency has ever been considered. Televerket behaves very much like a corporation, but its ties with the public sector are still very strong.

Hence, I want to emphasize the fact that for most European states, liberalizing the telecommunications market has not been equivalent to forming one or several privately owned telephone companies. The French situation is quite demonstrative of that behavior. This chapter outlines the historical background, summarizes the French government policies of the past few years, and suggests, in conclusion, my views on possible future moves and lists the major issues at stake in the immediate future.

AN HISTORICAL PERSPECTIVE: BEFORE 1974

From the first days of the telegraph, there has not been a formal monopoly on the telecommunications networks and services in France; when the first telephones were installed in the Paris area, more than 100 years ago, there were in fact installed by private companies operating with a license from the Ministry of the Interior: the first telephone service was opened around 1879, and it was based on the legal terms established 50 years before telegraphic service. France had then the equivalent of a general license system allowing companies to apply eventually for a license to operate the service.

Although nationalized in 1889 in a political context in which even business circles were demanding this, the telephone was still run under a general license agreement. The emerging national network was finally operated by the Postal Department, a part of the French Republican administration.

Kept in that same administrative position over most of the twentieth century, the French telephone system was unfortunately maintained in a state of underdevelopment because of its relatively secondary importance within the PTT: investment was notoriously underestimated, research was scarce, and service was rather poor. In the late 1960s, the situation was so rotten as to raise real trouble within the country, with libels being published and user groups protesting in Parliament. A recovery plan was started around 1968–69, and the manufacturing industry reinvested broadly under the incentive of research and development (R&D) government contracts by the early 1970s. A scheme was devised in 1971 to separate the PTT from the state administration and to form a public authority, similar to what was done with the British Post Office in 1972. However, the political climate of the early 1970s was not favorable to this project: the French President Pompidou was already ill and died in office in April 1974; the government had other priorities in economic and political terms, so the reform was never tabled in the French Parliament agenda at this time.

RECONSTRUCTING A MODERN NETWORK: 1974–85

However, the laboratories and the industry were at work, and the industrial background was getting ready for a boom: by 1974–75, the telephone became a real priority at the national level, with a visible will of the newly elected President Giscard d'Estaing to give all political backing necessary to restore the telephone network. From 7 million main lines in 1974, the French network grew quickly to 22 million main lines ten years later, and it is still growing today, with a more than 95 percent penetration ratio in French homes (28 million main lines in December 1990).

There were several major decisions made in the early 1970s to back that major development, industrially and financially. As far as the industry, France never had any vertical integration between the telephone administration operating the networks and the manufacturers: these were maintained in the private sector, as providers of the network equipments. This share of French manufacturing is still well and alive and fairly competitive today on the world scene (ALCATEL, SAGEM, MATRA, TRT, and other manufacturers).

As for finances, the French government has never subsidized the equipment of the networks. When money has been needed, the trick was to establish a bond system, to borrow money on the French and international markets, to finance the telephone development by private funding at market conditions. Started around 1970, this system is still alive, with a dozen financial companies dedicated to the investment in the network, who acted as providers of funds for the telephone extension, so that the French national budget has not been contributing to this equipment over the years. Instead, the revenues gained from these 15 years of big investments (1974–89) have flowed into the state budget at an accelerated rate since the late 1970s. Hence it should be stressed that the French Telephone has not been subsidized by the state but, on the contrary, that the telephone has steadily subsidized the French state! An account quoted in October 1990 in a public meeting by the Director General of France Telecom evaluated this overall contribution at more than 100 billion FF between 1982 and 1990.

In the meantime, the administration has opened new opportunities for private companies in several fields, including the following decisions that have marked the French situation over the past 20 years or so.

1. The terminal market has been progressively liberalized over the period 1960–86, with a wide tolerance that market providers can

sell modems, telexes, videos, personal computers (PCs), handsets, pagers, minitels, radiophones, faxes, scramblers, and so on, provided only that those terminal equipments are type-approved prior to their commercialization and installation. As a matter of fact, France Telecom is still a provider of terminals today, but a minor one for most of these equipments. Quite a few commentators were hence surprised by the French government position when a European Economic Community (EEC) directive liberalizing the European market for terminals was challenged by the French Republic!

2. The PABX market has been competitive and well for decades, and France has been, and continues to be, a very active market for all types of PABX, local and foreign, with a very tight commercial battle on prices, installation, and maintenance. The PTT has had a very small share of that competitive market, under 10 percent in present terms, whereas most European states had a strong monopoly until recently in these equipments (the United Kingdom prior to 1984, Germany until 1989, Holland until 1988, Spain and Italy until 1991 and probably still longer).

3. A great number of small- and medium-sized companies were formed either to act as agents for the major manufacturers or to install, to sell, and to maintain terminals and PABX. Many private small companies have been starting up office blocks, new housing developments, and so on. For some 15 years already, some of these "installers," as we call them, do subcontract also with the public administration to act in its name at various levels: it is not unusual in Paris or Marseilles that operations like wiring a new client, maintaining poles, or restoring lines are subcontracted to private firms by Telecom.

Taking advantage of the general license system established in 1837 and still alive until 1990, a certain number of licenses have also been issued by the French state to operate privately owned and privately operated circuits and systems. The major private networks were set up for railways, for power and electric supply, and for pipelines for gas and water supply. As a rising number of those private licensed systems use radio links of some sort, whether for data or for voice, there are hundreds of licenses for base stations; most of them, however, are not connected to the public switched systems and have been operated on a purely self operation by taxis, ambulances, maintenance trucks, and so on. This whole sector was maintained outside PTT operation.

4. Wireless and television stations have relied heavily on their proper infrastructures, operated also with a state license. From 1982, frequency allocations of a significant part in the spectrum were progressively liberalized, with hundreds of private FM stations

licensed over the country. Four television private networks were licensed in France, competing with three public networks of national audience.

5. Leased lines have been under a growing demand since the late 1970s. They are provided only by France Telecom to thousands of users, for all types of use, in a growing number of locations. As the national networks were modernized, it is now possible to rent these lines in almost any location in the country, with a month's notice at most. Most lines are digital, and they are still at a reasonable price. Most of the private networks do use leased line, even when they also have private dedicated infrastructures, like the railways do. There has been, however, some restriction on interconnection of private trunks with the public networks, and this has put on some pressure for regulatory adjustment since the 1980s.

In summary, when the Chirac government took office in April 1986, the French telecommunications scene had become fairly good, with fine territorial coverage, a high level of public satisfaction, a major industry, and a number of private companies operating with and around the state administration. Those firms were able to compete on an open market. No legal change in the PTT status quo had been in sight for many years, however, for the public opinion was not seriously demanding any change either in the management or in the operation of the telephone network. This was the framework in which an attempt to deregulate the PTT entered the picture in mid-1986.

A DRIVE TOWARD LIBERALIZATION: 1986–88

Under the conservative government that ruled over these two years, several significant moves toward liberalization of state-owned and state-managed enterprises were taken. Privatizing major banks and large national industrial companies was an objective of the Chirac government. PARISBAS, SOCIETE GENERALE, CGE, and many others were privatized. However, it was not considered then to privatize some or all of the PTT activities. The issues for PTT reform were explicitly stated as follows by this government policy.

1. The state-owned network was recognized as a vital infra-structure of the nation. As such, it should be run as a public service, with general interest and community service obligations. This implied maintaining the network as a major asset of the public sector.

2. Although no legal monopoly had been recognized for the national French operating telephone agency by law, this agency acted since 1989 as if it were guaranteed a monopoly position for most

telecommunications services. Some steps should, therefore, be taken to test periodically the efficiency of that monopoly, which shall be tolerated only as long as it functions well and serves growing demand for connections and services within the French republic.

3. Besides, competitive operations like radiotelephony and data services shall be driven more by commercial considerations than by public service obligations. Most of the telecommunications services should then really be considered as commercial ventures and be treated on a competitive basis.

4. With the international environment consistently requiring a freer flow of goods, services, and capital and with the EEC single market 1993 deadline approaching, the national telephone operator should become a credible and competitive investor on the international market and be able to invest and take risks anywhere outside France when it is worth it.

The above analysis was shared by most interested parties at the time by the government that inspired it, by the management of the French operator, and by most user organizations. There was, however — and there still is — a significant reluctance of the trade unions to consider the consequences of this analysis. Nonetheless, several decisions were made during that period, taking the previous assumptions into consideration. Let me summarize some of the more significant ones, because they have conditioned the more recent evolution of the French picture.

The first decision was to create a special name for the French operator, to baptize the network and form all visible patterns of an industrial group. "France Telecom" was the name publicized since the third term of 1987 and publicly advertised from January 1, 1988, with a significant campaign, national and international. For two years, this was nothing more than a brand name, because the operations were still totally administrative.

The second decision has been to rebalance drastically the telephone tariffs, taking care of the economic consequences of technological change in switching transmission, and use: time-sensitive pricing was introduced for local calls in 1986; the time parameter was quickly adjusted to reach three minutes per unit of account in 1988; international calls were made significantly cheaper on the major links (North America, neighboring countries, Japan); and discounts were also introduced for night periods and holidays, even at the international level.

A third decision was to separate explicitly regulation from operation by creating a distinct administrative body with an autonomous budget and staff. I created and headed that regulatory body from September 1986 to May 1989 outside of France Telecom, with the task of isolating the regulatory powers from the operating responsibilities of the network. Several major licenses were issued during this period, notably the license of a private radiotelephone operator incorporated late in 1987, and new networks have gained a national coverage since 1990.

Finally, an attempt was made to produce a bill proposing to transform France Telecom into a commercial company whose shares would have been held mainly by the state, with shares sold to the employees. The personnel would have to quit the public administration over a period of two years. However, this bill was not introduced in Parliament, and no statutory change was made prior to the change of government in April 1988. The public debate was opened, however, and was going to continue for two more years prior to the enactment of a PTT law in July 1990.

CONCENTRATING ON THE PUBLIC SERVICE: 1988–90

Following the presidential elections in spring 1988, the French Parliament elected in spring 1988 had no more conservative majority. This government, led by M. Rocard until May 1991, immediately rejected any new step toward the formation of a PTT commercial venture; all moves toward privatizing public bodies have been stopped since then. The new policy lines were thus to enhance the values of public service organizations as they were traditionally recognized in the French administration for more than three centuries! Although the previous Post & Telecommunication Minister, Gerard Longuet, had built up his policy around the idea that competition would have a positive effect on France Telecom, his follower, Paul Quiles, considered and declared immediately after entering office that no move of the French government should aggravate the impact of international deregulation on the French public telephone service.

This shift in policy struck international observers: much has been said and written since late 1988 in the trade press and newsletters about the so-called conservative or monopolistic views of the present rulers of France. I think, however, that a more realistic approach should be taken for practical day-to-day operations of France Telecom and for the rest of the industry. However, the climate had changed, and the emphasis was deliberately put on a central and monopolistic role of the PTT for all communication services, present and future.

New commercial ventures and partnerships have nonetheless been opened in France during this period, notably with respect to international value-added services, foreign investments, software development, and so on. Continuity was particularly clear in the field of tariffs.

No change was made in the tariff policy, although the reorganization of telephone pricing has been stopped recently, with no more reassessment of local and long-distance tarification and little change in the leased lines pricing scheme.

The PABX terminals and services markets were maintained much more open in France than in any similar country in Europe; compared with Britain, Germany, Spain, or Holland, these markets are still more competitive in France. The share taken by the national operator is, by far, smaller in France than in any other country in Europe.

There was no modification of the licensing procedure until the end of December 1990. The existing licensees continued their operation as before either for their own sake or for commercial use of those given licenses for value-added services and telematique services.

In the meantime, the European Council of Ministers has taken two steps toward the integration of telecommunication services and networks in the EEC. These two moves were implemented by France and its 11 partners.

The liberalization of the terminals market was decided with a directive of the commission issued in May 1988, a second one being drafted in 1990, to organize the mutual recognition of type approval among member states. The liberalization of services and the recognition of international standards in this field have started with the notification in July 1990 of a "service directive" of an "open networks directive," which should both be implemented by the member states over a period of five to six years at most.

Several tough periods have been encountered in this European frame since 1986, all due to diverging views among the EEC member states about the future organization of the unified Common Market. However, harmonization of practices, if not of tariffs and conditions, is underway in Europe, and the French policy has been consistently in favor of this harmonization despite the changes of parliamentary majority and the change in governments in 1988. Also, no one in France really dares to suggest that France should step back from the implementation of the EEC Common Market, neither for the postal nor for the telecommunications services and equipments.

The main divergences between the present French government policy and some of our partners in Europe refer to the respective roles of the state-owned undertakings versus privately owned companies and to the priority given to each of these vested interests in organizing the European Common Market as well as to the degree of open competition tolerated or organized for the telecommunications networks and services. At present, the French government keyword is "organizing the market," while, for instance, the key phrase of the British government would be "as much competition as possible" for the sake of the user!

That image of the coming French market for telecommunications as a whole does not, however, give full account of what has been really different in this country when one compares it with similar countries in Europe or elsewhere; as a matter of fact, the telematique program has highly conditioned the approach of new services in France and thus is worth special attention.

The "telematique" concept dates back to 1978, when a French government report used it to qualify the overlap between computers and telecommunications. Initially, the French program was designed to provide electronic services to the homes as a complement to the telephone network. Through the free distribution of Minitels (videotex monitors), a handy billing system, and a series of good guesses, French telematique became a world-unique success story. It involves now so widely the French society as to be included in the "French way of living."

THE FRENCH TELEMATIQUE EXPERIENCE

As recalled above, when elected in May 1974, the French President Giscard d'Estaing found the French telephone network in poor shape: one of the lower penetration rates of the industrialized world, an average wait of four years to get access to the network, and most rural areas in the country still equipped with manual switches. Prepared for some years in the back office, the drastic reform of the telephone was strongly supported by the president, who made the telephone a national priority. Underlying this political choice, three policy lines were implemented during the following period (1974–78):

1. The industrial level. Because most trunks and switches existing at the time were outmoded or exhausted, the decision was made to implement a digital network as soon as possible and as early as industrially possible. This accelerated the industrial production of digital switches and transmission equipments and allowed quick planning of a packet switched network intertwined with the restored telephone network and tariffed on a distance-insensitive basis.

2. The marketing level. Because a majority of the French people had not yet used a telephone set at home, it was considered feasible to mobilize all energies to introduce new information technologies through the telephone line and to diffuse digital technologies in the French society at large. This was the rationale for Nora and Minc's report; "telematique" was first used as a forged word in this famous report handed out to the French president in 1978.[1] The new concept was the heart of this report, which proposed a public policy to develop information services alongside the newly restored telephone network.

3. The societal level. With these goals in mind, a growing majority of the public became worried about the inherent risks to a computerized society, such as infringements of individual privacy, protection of human rights against undue scrutiny, and

administrative procedures handled by computers. Therefore, a strong involvement of the president himself into seminal studies and public debates[2] centered on informing French society and the January 5, 1978, statute on "informatique and libertes," the French data protection legislation, was enacted.

Telematique was ready to serve as a locomotive pulling new services and new social behaviors behind it! To the telecommunications community, concentrated around its chief executive, Gerard Thery, then director general of the public telecommunication administration in the PTT department, telematique served also as a leading project able to mobilize all energies within the operating administration and around it. The main selling argument for this program was the electronic directory, complementing the phone book and eventually able to substitute for it in the long run. It was an easily accessible service, handy to use, free of charge, carrying ads, and a fantastic tool to promote long lines of traffic, for any customer could then find the number of any remote correspondent without the usual chase through an uncertain and costly operator service.

The free distribution of the Minitel terminal was linked to that perspective. This was supported by two strong ideas: electronic emulation of the phone book, which has never been charged by the telephone service, and the hope to pay for the new equipment by the traffic flow it was supposed to generate after a few years. This tiny, simple equipment was bound to become an annex of the telephone set and to live up to a "network effect," just as the simple old telephone did when it was first introduced some hundred years ago.[3]

It should be remembered, however, that the French telematique program did not have a simple start. When the Minitel was introduced in late 1978, a significant share of the newspapers called for a stop of this program. Worth mentioning are declarations published by the press of the time, like this one: "New modes of information will threaten the liberty of each citizen!"[4]

To cope with this apocalyptic forecast and to get back to real data, a large experiment was designed in the Paris suburban area, including a sample population in the Versailles area. The code name of this operation was Teletel 3V. On purpose, a few regional news papers were involved in this experimental test phase. A socially representative sample of the population was drawn, counting 1,500 homes, plus an oversample of 1,000 to complete the study. All homes received a free Minitel, and many professionals were invited to join, either on the provider side or on the analytic side, with quite a few social workers, economists, and lawyers involved in this on-line survey.

Soon, 120 organizations joined the program, all of them being sponsored by the operator, in order to design and test a variety of services that became available to the 2,500 sample. After two years, some clear-cut data were available:

A good third of the sample never used their Minitel. (This was consistent with the longer-term experience we have acquired since then.) It meant that this share of the population was technoresistant to the innovation.

Twenty-five percent of the sample became Minitel addicts, carrying more than 60 percent of the traffic generated on the network. They were the proselytes of the system.

When actively involved, household members did appreciate the interactivity of the new services, quick updating, games, and "chat" lines.

From this experimental phase, the press finally got involved heavily in the service business: they had witnessed the fact that good money was available there and that it was not generating any competition to their traditional market base but was opening new windows in the service business; today, not a single newspaper can live without its own telematique service, whether it is "Le Monde" or a regional paper!

So the program was finally launched on a national level with a fast-growing number of terminals and services available in the country: 120,000 Minitel in 1983, with 140 services on-line at the end of this year; it grew steadily year after year to 3 million Minitel in 1987, when more than 5,000 services were accessible, and reached 5 million terminals in 1990, with approximately 12,000 services registered on-line.[5]

KEY FEATURES OF THE FRENCH SYSTEM

Impressive as they are, the above figures do not reveal all the tricks that made this success possible, and the significant throughput of the French videotext system should have some hidden cause that made it different from its competitors established in other countries. As a matter of fact there were, in my opinion, five significant features of the French experience that jointly created favorable conditions for a success story. Most of these features were not built into the other videotext services like Prestel in England, Bildshirmtext in Germany, and Telidon in Canada. Teletel was significantly original on the following points.

1. The terminal was the most visible item. It was designed to be simple to operate, simple to manufacture, and simple to install and to remove, as a visible appendix to the telephone set. From this design grew the idea to give the terminal free and pay for it from the amount

of telephone traffic it may generate. Because it was small, free, and easy to use by everybody, it soon became a popular home device, like the phone and the television set. This was an option very different from the one made in the United Kingdom, where the videotext installment supposed a cross connection between the telephone and a television set, and different also from the U.S. approach of extending home PCs toward videotext for the computer addicts! The French way appeared to be the most natural one for large public usage.

2. Then came the second trick: to define, market, and tariff Teletel services as an extrapolation of the telephone service itself. The operator never considered the Minitel a cheap computer device; on the contrary, it was treated as an enhancement of the telephone.[6] Every effort was made to stimulate this friendly, user-oriented tactic, on the terminal itself as on the service level. The natural language was prevalent in all applications; there was absolutely nothing reminding the user of the computer. This approach was consistent with the fact that the terminal was brought to the public by the telephone administration, a state operator, and it was also consistent with the friendly, anonymous billing system installed from 1983 onward, with the strong support of Jacques Dondoux, then director general of the telephone administration.

3. This billing device was also an example of the originality of the French network: baptized from the casual name of the newsstands — "le kiosque" — this billing has proven to be ingenious, simple, and very cheap to handle. It cuts down all administrative costs incurred in listing subscriptions (which are no longer useful), billing individual clients, handling small payments for scarce service use among a widely dispersed population, following each account individually, maintaining the unsustainable carriage of passwords for a large population, and so on. Everything was delegated to the telephone bill: all consumption was billed on a timely basis, leaving it to the telephone operator to pay back their due to the service providers, according to the amount of time consumed by the callers of each service. No fee was charged in advance to anyone, except to service providers for this packet switched network entity (however, this professional-to-professional relationship did not require a great amount of administrative cost). With this system, the consumers have all available services accessible at their fingertips; they may anonymously use them daily or yearly without any formality. They have guaranteed privacy, for no one is able to trace users' habits, unless they are willing to declare their identity on purpose to the service provider. When a train reservation is made, for instance, the phone billing becomes as discrete as the newsstand where one buys a journal, and payments are fungible under the general heading "Minitel use" on the phone bill. This has certainly exacerbated the use of "pink Minitel," chat/porno services about which many groups

were worried. However, it also allowed useful developments at almost no administrative cost and the start-up of many successful service operators now ranked among the 20 leading computer service firms in France. There has not been any equivalent boom in the world with another videotext service.

4. A clever engineering device helped to make Teletel different from its neighbors: because the flow of information derived from each terminal is very small indeed, by construction, to stay user-friendly and because people do not usually type fast, the design engineers decided to use the PSN Transpac as the main carrier for any distance carriage while the telephone network carried data from the local exchange to the user. Local lines were tariffed cheaply, and Transpac pricing was proportional to the flow of bits, whether it stays distance insensitive. Service provision became independent of the location of the caller as well as independent of the location of the service computed. Any service provider thus had an equal chance to gain a national audience; in practice, the average carriage cost dropped from 200 FF per hour for a long distance call to 20 FF per hour for a typical service provider. This interconnectivity made a fantastic economic difference with all other videotext tarification then enforced and allowed vast spreading of the service business.[7]

5. Finally, service provision has been free of any operator involvement except the electronic directory service, which, of course, was handled directly by the administration. The service creation and development were totally market driven as soon as the Teletel 3V period was over. This liberal, decentralized, open policy was opposite to the philosophy followed in England by Prestel, where all files were supposed to rely on British Telecom (BT) computers, as was also the case for the BlX and Videotel and most other videotext systems elsewhere. This voluntary French open policy was chosen on purpose to be consistent with the Transpac decentralized architecture, as explained above; to avoid any monopolistic position from the administration, for such a position would have raised again a furious criticism from the press lobbies; to allow a flat of service dependless of who provides the service and where it comes from in the country. This policy did demonstrate that the public operator was not involved in service definition and provision and made Teletel one of the most explosive markets ever experienced in the country. Teletel protocols were instrumental in this sense, for they made it possible to create a networked flow of bits connecting any phone subscriber to any service provider in the country at a flat rate.

RECENT DEVELOPMENTS

More recently, billing was diversified on the kiosk, mainly to allow a wider spectrum of prices for the sake of professional services.

Services remain accessible through simple mnemonic phone numbers in the four-digit series from 3613 to 3622, with eight levels of possible pricing. Flat rates remain the principle, but services can be billed anywhere between 0.13 FF.mn. (3613 services) and 3.65 FF.mn. (3619 international services). All such services are billed through the telephone bill of the user.

As a whole, the success story is now behind us. New questions arise linked basically to two problems: Will we ever significantly cross the French border? Despite the precautions taken from the origin to maintain videotext services in a free market, can we get rid of the dominant position endorsed over the years by the telephone administration to run most of the Teletel back office?

The French system has not grown internationally. Many explanations are given, but none gives a complete account of this failure. The Minitel story was, I believe, a very French one: it was handled as a national system, consistent with the peculiar history of the French network, related its poor shape in the late 1960s to the voluntary push given by the state, and sustained by political will throughout the 1970s and 1980s. Very little of this prepared Teletel for an international competitive involvement. On top of this, the significant competition in design, standards, and protocols all around Europe made it ethnically difficult to explore other territories until now. It would have been impossible for a French public operator to compete visibly with its neighbors on their own territory without a reciprocal provision or an exchange of market entry. The only way to go international has been to escape any direct confrontation with the other European operating agencies, either by a compromise or by a third-party entry. Also, pricing of transborder services remains too high to gain a significant audience, and handling of international videotext is still technically difficult and troublesome.

There are, however, reasons to think that the current trend toward a freer flow of services within the EEC may open new opportunities for international service provision. It has started with the bilateral agreement between France and Germany to cross borders through the kiosk 3622, specifically designed to facilitate Teletel/BTX services. At the same time, newly compatible protocols are accepted to ease international flow; however, we are still at the starting point for that effort.[8] As a commercially oriented policy makes its way through the European states, both within the EEC and around this cluster of nations, there is a growing chance to see multinational firms getting involved in Teletel services and to ask for an international extension of these services (for example, the German automobile manufacturer BMW, whose service is very efficient in France, provides support and management tools to their dealers). Those firms need to extend their facilities abroad. A more liberal policy in Europe, a growing commercial interest, and a good marketing strategy of the operators will certainly do it. When? We shall see, hoping we do not wait too long.

Teletel was born from France Telecom, still called, in the early 1980s "la DGT." It is now a grown-up child, almost an adult, able to take care of its own future and to behave independently. As a mature industry, Teletel also is now ready to support a market-driven competitive regime in France and abroad. This, of course, means that the parents, France Telecom and Transpac in the present case, are bound to let their child grow independent. Certainly, this maturation is not simple to live through. One of the magic tricks mentioned in this chapter, the billing device, is at stake in this process: quite a few interest groups have already asked to bring the billing system back into a competitive regime. They suggest to form alternative billing services, competing with the telephone bill, in order to bring themselves a portion of the money flowing into the system. Impossible at the start-up period, when subscriptions were hazardous, this scheme now becomes feasible. Press groups, banks, and mail-order companies are ready to step in there and to compete with the telephone kiosk. Will these newcomers be allowed to do so? Nothing is clear yet under the present regulatory system, which gives an arbitration power to the state operator, just because it is still an administration. Since January 2, 1991, France Telecom has been a national agency, somewhat similar to what BT was in the period 1972–84, before its privatization as a public company. Will this be enough to change France Telecom behavior and to allow a competitive entry on such a sensitive factor as the telephone bill? Two factors will enter the picture, in my opinion: the interpretation given in practice to the recently confirmed telephone monopoly claimed by the French government in favor of France Telecom,[9] and the interpretation given to the service directive published in the spring of 1991 by the EEC Commission.[10]

In the meantime, 12,000 service providers make some 6 billion FF per year in serving 5 million customers on-line. Even on a national basis, this is not too bad a picture.

THE FRENCH PTT REFORM: SPRING 1989–91

The open "public debate" driven by a government task force in December 1988 was a typical trend of the Rocard government period (May 1988–May 1991); the job was led by Hubert Prevot, a close friend of the prime minister, formerly a leading staff member of a trade union. His task force opened a wide series of public meetings, calling both PTT and non-PTT witnesses to exchange views on the PTT's future.

The main purpose of this method clearly has been to pass on the message "your future will be different from your past, but, do not worry, you will stay in the public service, if you wish, and your role in society will still be central." Among the various attempts made by this late Rocard government to refurbish the values of state public services, the one in the PTT was probably achieved the most and was the most successful during this three-year period, for its final outcome was the enactment of the law revising the statutes of the Post and of France Telecom.

This draft law was made public during the winter 1989–90. The bill was threefold, was basically voted without major changes, and was enacted July 2, 1990.

First, two separate public operators were formed, one (La Poste) in charge of the mailing operations, the other (France Telecom) in charge of the telecommunications networks. Each of them incorporate the assets, the personnel, and the client base of the former branches of administration for the PTT. The two agencies remain totally under public state control, and neither takes the form of a corporation. The agencies have separate assets and separate accounts, but no shares and no capital.

Second, the former workforce of public servants remains in the state public service, and their employment is still fully in line with the statutory rules of the civil service: recruitment, careers, social benefits, wages, and full life employment are maintained. From this standpoint, there are no significant changes in personnel management and very few (or minor) adjustments made on the industrial relations of the PTT.

Third, the PTT administration per se is restrained to the most administrative and regulatory role, with two distinct bodies established in the ministerial department in charge of the PTT. One is called "regulatory Directorate," and it takes over the previous Regulatory Task Force I had established in 1986 to monitor the private sector, to administer frequency allocation and type approval, and so on. The other is called "public service directorate," and it regulates the public operators, La Poste and France Telecom, while both "regulators" report to the same ministerial position in the government.

This organizational reform of the PTT became operational in January 1991. On the positive side, it consolidates the informal separation of P&T that had taken place over the last quarter of the twentieth century: the two operational branches of the PTT have had little, if any, cross subsidy over the past decades in France; recruitment was already distinct for many years; accounts and managements were practically independent from each other for decades, at least since the 1960s; investment programs were fully separated since 1923, when separate, independent budgetary lines were established by law for the mail and for the telephone.

On the positive side, also, is the fact that the contractual links with the public become now closer to corporate law: commercial contracts begin to be the ruling reference with third parties, whereas the former administration had to deal with the administrative law and administrative courts for all matters, including consumer relations and services.

One could still consider that maintaining the whole personnel structure under the overall umbrella of state employment is also a positive side of this reform, for it has frozen most union claims that dealt with a possible privatization of the PTT employment. This analysis has led part of the political opposition, headed by former Premier Raymond Barre, to favor this move, despite the views that this legal reform was unable to cope with the international deregulated environment that is restraining, year after year, the closed fields of a national public service like the PTT used to be, decades ago.

Nonetheless, the French PTT reform of July 1990 was slow and probably inefficient on several important points. Probably the more critical item is the international credibility and appraisal of this reform. Business-wise, La Post and France Telecom are clearly maintained under the influence and control of the French government; their investment programs are fully controlled by the Treasury and result from continuous political bargaining at government level. As an example, France Telecom involvements in cable television distribution, in a satellite program like TDFl, and in advanced high-definition television standards are not formed according to a corporate analysis but pertain to the program of the leading political appointees in the French government. Such decisions are dealt with in the president's office and are discussed with the parliament majority. This visible political dependence goes much beyond the long-term investments program quoted above: even daily operation or tariff announcements are made by the former minister (as they also were announced by him before the July reform) and not by the chairman of France Telecom, who has restrained his public appearances to less-public announcements. Similarly, the French minister of the PTT gets involved on the front row of international negotiations with countries like Mexico and Argentina, as if he were the actual chairman of the French national operator (whether the ministerial position may rather be the one of a referee than a player on the field!).

Although acting according to a model closer to a regular corporation than to "la Poste," France Telecom thus stays under very strict control of the French state. It is noticeable, for instance, that the government formed on May 15, 1991, by the new Premier Cresson (the first lady to take that post in France) puts the PTT minister under the umbrella of the Secretary of State of Economy and Finance.

This government structure raised many eyebrows because over the years, the main preoccupation of France Telecom management has been to escape as much as possible from direct reporting to the French treasury for fear of maintaining a strict administrative control by the government of the high returns from France Telecom (6–8 billion FF per year).

Another point at stake after the July 1990 reform is the ability of the public operators to adjust their behavior to market demand and to tolerate competition as the normal way of doing business in their fields of activity.

FUTURE TRENDS AND ISSUES AT STAKE: 1992 AND AFTER

With the European and worldwide environments moving fast, one notes that the rules are being modified, step by step, to integrate the moving patterns of technical supplies and economic demand for telecommunications products and services. Until now, the French government policy has been inspired, in the long run, by a "middle of the road" pragmatism: no privatization, not much internal competition on the operating side. At second sight, however, something might be moving, but at a prudent and slow pace. The "public debate" opened during the first half of 1989, whose purpose was officially to check what the public and the PTT personnel thought about their future role and operations, has helped to raise the consciousness of PTT employees on the necessity to change their long-term views.

The summary of this consultative national forum was published late August 1989. It has, of course, raised a series of criticisms from quite a few union leaders. Nonetheless, the government had talked with the personnel representatives to check the feasibility of reforming the PTT structures and the statutory responsibilities. This was the start-up of a process that over several months included implementation of structural adjustments under the pressure of European integration.

A first adjustment dealt with international matters: France Telecom has both opened international offices abroad — in the United States, in Germany, in Japan, in the United Kingdom for instance — and set up new ventures with other operators and industrial partners. The most significant of these ventures are a 10 percent stake in the UK operator licensed in early 1990 for CT2-Telepoint, a 50 percent stake in the German value-added services consortium started jointly with the German Bundespost, and a 15 percent stake in the international data service operator infonet.

A second adjustment concerned R&D. The French operating agency moved slowly toward a different view on R&D. Although traditional programs were considered on a purely national basis, in

France and abroad as well, new concepts were tested on emerging protects: the future GSM generation of cellular radio systems was developed with international partners, trading with the various operators in Europe. Therefore, France Telecom R&D contracts concur to this development with specifications common to the Bundespost, the Dutch Telecom, and others like Telefonica, SIP, and BT.

A third concerns the openings of cross, international ownership of new activities with other operators France Telecom bid unsuccessfully for a stake in the Chilean Telco ENTEL but was more successful for Mexico, Argentina, and, possibly, Hungary. At the same time, BT and AT&T are opening business in France for telecommunication-related operations network management.

Many more examples could be quoted relating to the emerging multinational stakes in which the French operating agency is now involved. All those signs confirm that the French operator is becoming more international, more competitive, and more commercially driven.

Despite its present structure past the 1990 reform, France Telecom the administrative agency of the French Republic have been keen to adjust to this new international environment. Similarly, the exploding demand for diversified communication services appeals to new forms of investment, in which the French operator has taken shares.

The economic analysis of the trend is well at hand, but we were, until recently, short of real social acceptance of a quick move toward a quasicommercial body taking over the traditional tasks that France Telecom performed but accepting also the risky business in a competitive environment. The public discussion opened in the country since the operations have changed into a commercial public venture can now be considered seriously. My forecast is that this change is necessary but that it shall require still several years to be completed and efficient. In the meantime, France Telecom policy will continue seizing opportunities when they are available, but without any change either in the capital structure or in the social relations within France Telecom.

In other words, my view of the future is one of a creeping liberalization of the commercial telecommunication services continuing in France over the coming years, as it is already, I think, with the terminals market and mobile units. I do not see, however, any chance of quick privatization for the French operator, which was not considered either by the present government or by the previous ones.

NOTES

1. Nora Simon and Allain Minc (1978) *L'Informatisation de la Society* Paris Doc. France (Translated into English Cambridge, MIT Press, 1979)

2. TRICOT Report, (1975), Informatique and Liberties (1975) Doc. France, Paris, Informatique and Society, Acts of the Colloque de Paris, (1980) Doc. France, Paris (May all be referred).

3. The French Telephone started operations in 1879.

4. La Federation de la Presse Francaise (1978).

5. Marchand Marie (1987): *La Grande Aventure du Minitel* (English Translation: The Minitel Saga 1988)

6. Profit Alain (1990) *Le Communicateur* 11:75–78, Paris.

7. Dondoux Jacques (1990) *Le Communicateur* 11:19–21, Paris.

8. Maury Jean-Paul (1990) *Le Communicateur* 11:101–02,

9. Parliamentary Proceedings France on the Regulation of Telecommunications (1990, September 19) enacted into Law on December 29, 1990.

10. Draft for Deregulation of Telecommunications was sent to the member states of the European Community and released in final form in January 1990.

3

Deutsche Telekom's Strategy: Cooperative Communications Policies

Klaus Grewlich

Reorganizing its internal structure, rebuilding the communications network in Eastern Germany, and expanding its presence in international markets on the basis of cooperative arrangements are the three major challenges that Deutsche Telekom has to face today.

In Germany, 1991 marked two anniversaries in the telecommunications business: 110 years of public telephone service and the first year of Telekom as an independent company. With annual sales totaling DM 47 billion and with some 260,000 employees, Deutsche Telekom is Europe's largest telecommunications company. In 1990, Deutsche Telekom was separated from the Postal Administration and began to operate independently, facing competition and actively addressing the needs and wishes of its customers. Deutsche Telekom is now a company in a highly competitive market. Cellular communications and satellite and data network services, for example, are now open to competition. At the same time, Deutsche Telekom is developing a more international outlook.

GLOBAL ASPECTS

In the geopolitical sphere, the world that the international community has grown accustomed to is undergoing fundamental change. The clash of ideologies, the confrontation between military groups, tyranny, and economic stagnation are increasingly being replaced by democracy and the free market.

As a result of the globalization of activities in the economic sphere, enterprises are breaking new ground. Meanwhile, electronic networks are revolutionizing the way markets function. Cross-border mergers and acquisitions, worldwide procurement, the development of international joint ventures, cooperation among competitors to establish common standards, strategic alliances allowing competition to continue via selective cooperation, joint research and innovation programs designed to push back the frontiers of common knowledge — all these are becoming well-known developments. In view of this globalization process, special efforts will have to be made to establish cooperative communications strategies.

Traditional telephone companies become fascinated by the "global customer." Who is the global customer? He or she is a sophisticated user who knows where to shop around and is likely to know where the best deal is available. He or she might be a communication-intensive individual or a small company, but normally this customer is one of the thousands of truly transnational enterprises on the list of *Fortune* magazine. These "Fortune enterprises" today control approximately two-thirds of the world's productive capacity across five continents. They are the major customers of telecommunication operators and account for around 60 percent of operators' revenues today.

What do global users primarily need? They require a greater selection of communications equipment and services, free flow of information across the borders, efficient and timely standards, fewer regional differences with regard to available facilities, the encouragement of innovation, and skilled and committed personnel. Carrier and service providers will have to show a high degree of user orientation to successfully meet all these requirements, particularly in terms of facilities and services.

The challenges concerning investment and innovation are important. Data traffic volumes are expected to grow by more than 20 percent per year; voice traffic, on the other hand, is expected to grow by only 8 percent. Experts estimate that by the end of the twentieth century, a substantial part of business communications could well be in the form of data traffic, followed by video/image transfers, with the rest — still important — continuing to be voice communications.

Thus, the communications industry is starting to become globalized as international business becomes even more global. Regulation follows and promotes this process at the same time. The telephone companies — previously monopolies in their national markets — now find themselves competing with their fellow telephone companies to provide end-to-end services for the global customer. At the same time (and this is the really interesting aspect), they must cooperate to gain access to each other's network to be able to bring the best selection of services to the customers. This mixture of competition and cooperation is quite stimulating in a situation in which telecommunications is not just a piece of useful infrastructure but a make-or-break ticket to economic, cultural, and political development.

The computer industry, long since an integral part of communications and itself globalized earlier, is a vital partner in the globalization of networks. The broadcasting industry, already profoundly influencing international political and economic change as broadcast footprints reach across borders, is likely to be involved in a global network as fiberoptics begin to carry video.

Thus, national boundaries are becoming increasingly more artificial for global customers and operators and for service providers. These customers simply do not want to assume the burden of trying to "coordinate" the activities of dozens of different national suppliers. They want full service, single-end billing, and account management on the basis of one interface. They want a global supplier or a global strategic alliance that can design and manage the network wherever they need to go. If the established telephone companies are not able to do this job, customers might eventually do it themselves.

Where can we see the first signs of these global actors and global strategic alliances emerging in the field of advanced communications? At present, there are only bits and pieces — we are just at the beginning. However, what happens now might be decisive for the future. There is, of course, the development of broadband applications — broadband Integrated Services Digital Network (ISDN) — a very ambitious evolutionary process that will go on for a long time. On the other hand, there is a trend toward global data networks. You have heard about multilateral one-stop shopping agreements; there are other more recent and more focused efforts in that direction, going beyond just a collection of "one-stop shopping agreements." In addition to the truly global networks that are emerging now, there are regional network management and control projects, for instance in Europe, in the North American region, and on the Pacific rim.

Global networking is concerned not only with transport but also with advanced services that depend increasingly on sophisticated signalling and control technologies. Many experts are of the opinion that this is where the competitive edge will be found in the future.

Power has always been conferred on those who have information and knowledge and know how to use it. The dissemination of information technology can be a precursor to shifts in, or consolidation of, both political and economic power structures at national, regional, and global levels. The information revolution is changing the international economic system, transforming national, political, and business institutions, and thus is affecting not only the nature of national sovereignty but also the relative strengths of the regions and political and economic actors in world affairs. Although "communications" is a very competitive business, it will not be possible for one country, region, or economic actor to derive benefits from this development toward the "information society" unless it has the cooperation of others. This simple and central fact has to be kept constantly in mind. Cooperative communications policies are required.

Whereas international law was simply a law of international coexistence in the past, defining areas of sovereignty of the individual states is increasingly being replaced.

DEVELOPMENTS IN EUROPE

The European Community is actively participating in the emerging cooperative communications policies. What is happening in Brussels is clearly important. With regard to telecommunications, "Brussels" is several things at the same time: it is international trade policy, it is competition policy, and it is still industrial and technology policy. In the early 1980s, Viscount Davignon forcefully called public attention to the extent to which Europe lagged behind in the whole field of communications. At that time, it was customary for conference spokesmen to describe competition with the United States and Japan as a "rat race" — the United States and Japan being ahead and Europe struggling behind. This served certain purposes, but with international mergers and global alliances making the news columns every month and with multinational companies predominantly involved in several regions of the "triad," the clear-cut picture becomes complete.

Good economic performance is, I believe, an optimal way to hedge against protectionism, and liberalization and competition may be described as the best industrial policy to achieve good economic performance. Still, it might be reasonable for the Community to engage in joint research and developments projects — to promote, for instance, neutral standardization — where such policies prove to be efficient and are compatible with international economic relations.

What has become even more interesting, in the meantime, is the role of the European Community in the field of establishing the rules of the game. Telecommunications reforms continue to be strongly influenced by the blueprint for the future market structure set forth in the 1987 Green Paper and subsequent community legislation. The principles laid out in the Green Paper can be encapsulated as liberalization and harmonization. *Harmonization* means defining a minimum set of common rules for the entire Community that allow a single market in telecommunications to function. This is the way to secure the integration and interconnection of networks and services throughout Europe, to help the efforts of the telecommunications administrations who are the primary investors in the network, and to establish a setting in which service operators can provide Europe-wide services over the network and users can connect the terminal equipment they wish to use. The accelerated pace of growth for networks and services means that manufacturers will be facing a growing demand for their equipment. *Liberalization* fosters the competitive environment that alone can make a new range of equipment and services available at affordable prices. It requires the separation of the referees from the players; it means ensuring that commercial interests operate separately from regulation and government.

The overall approach rests on establishing the right policy mix, an equilibrium between the introduction of competition in as many areas of telecommunications (terminal equipment, non-voice services, and so on) as possible and the acceptance that other "reserved" areas (network infrastructure and voice telephony) will not yet be open to competition.

Telecommunication services are now a 100 billion ECU market in the Community. June 1990 saw the final adoption of the Open Network Provision (ONP) Framework Directive that defines the guidelines for opening and harmonizing the conditions under which new service providers and users can gain access to the network infrastructure. The Commission's Directive on competition in telecommunication services, abolishing special rights of telephone companies (except voice and network infrastructure), and the ONP Framework Directive came into force simultaneously on July 1, 1990.

Toward the end of 1990, the Commission published a Green Paper on satellite communications, inviting public comment. The satellite Green Paper aims at a fundamental reform and liberalization of this sector. It seems that the Commission will soon publish a Green Paper on an overall approach to mobile communications. What is the attitude of the operators in view of the developments underway in Brussels? Public operators recognize that competition in telecommunications is the rule, not only in the United Kingdom and in the Federal Republic of Germany but also in numerous other countries in the Community. Private companies are now offering every kind of terminal or service (except for the telephone service) in competition with the established telephone companies.

The growing complexity of the technical systems, the accelerated pace of the cycle of innovation, and the increasingly more sophisticated demands being made by our customers make it necessary for us telephone companies not only to strengthen our performance as we compete against each other but also, at the same time, to coordinate some of our actions to an even greater extent. Only in this way will we be able to prevent the fragmentation of the Europe-wide and worldwide telecommunications systems and not put this instrument at risk. Another initiative along these same lines is the establishment of the European Telecommunications Network Operators Organization (ETNO). We view ETNO as the umbrella organization that will build on the resources of the newly specialized institutions such as the European Research and Strategic Studies in Communications (EURESCOM).

It might be appropriate to address the issue of the overall responsibility European network operators are supposed to have for an efficient European telecommunications equipment industry. Although I do not believe that European operators should have particular responsibilities for the European telecommunications manufacturers — it is plain economic reasoning that operators procure worldwide wherever quality is best and prices are optimal — it is nevertheless in the interest of the operators that worldwide competition is functioning, and that implies also the vitality and efficiency of the domestic European industry.

Telecommunications operators should take into account the need for a homogeneous European market, including international standardization. Such European efforts — that means the preparations for the 1993 Single market — do not necessarily conflict with Atlantic initiatives, with efforts within the Japan–U.S.–Europe triad, or with future work within the CSCE framework. One example is the European Telecommunications Standardization Institute (ETSI), in which the non–European Community Organization for Economic Cooperation and Development (OECD) countries are participating. Another example is the Standardization Summit held some time ago in Frederiksburg, where the European Standardization Institute ETSI, the U.S. top standardization body T1, and the Japanese-led Pacific ISDN Council worked together. This effort is not a government summit, but a process promoted by enterprises. It is not enough to focus just on Western Europe, that is, the European Community. Telecommunications have been instrumental in the process of political change occurring in Eastern Europe and the former Soviet Union. An OECD study has described communications technologies as "freedom technologies." Now, communications are vital for sustaining the democratization process and for economic recovery in Eastern Europe.

It is a major task to open up and make adequate tools available in the support of efforts to develop telecommunications in Eastern Europe. In the Eastern part of Germany we can observe how challenging it is to establish the physical basis for interconnection and access to networks and markets. I believe that from the point of view of the operators, two approaches are necessary.

First, Eastern European countries must be fully integrated into Western European telecommunications institutions such as the Conference on European Post and Telecommunications (CEPT), the newly created European Telecommunications Network Operators (ETNO), the ETSI, and EURESCOM. At the same time, the instruments provided by the Community such as PHARE, as well as loans from the EIB, the European Bank for Rehabilitation and Development (EBRD), and the World Bank, must be used to stimulate the introduction of telecommunications.

Second, operators should advocate that more be done; a wait-and-see attitude would not be sufficient. We have to make a substantial contribution to telecommunications in Eastern European countries now, and this is not motivated just by a desire to make short-term "quick-kill" profits but rather must be seen in a long-term perspective, in which market presence, market development, and global positioning are important. The European Community should consider making the Community's appropriate tools available to support efforts to develop telecommunications in Eastern Europe. European operators might go into this market — in order to share know-how, money, and manpower — in varying alliances and consortia that would be open to non-European partners, an interesting and potentially beneficial mixture of competition and cooperation.

CHALLENGES AND STRATEGIES

Let me now come to the strategy of Deutsche Telekom within the framework of the telecommunications reforms in Germany.

Regulatory Telecommunications

The regulatory telecommunications environment in the Federal Republic of Germany has the following key elements:

First, the political and regulatory tasks have been separated from the operational and entrepreneurial ones, whereby the former, such as type approval of telecommunications terminal equipment, are the responsibility of the Federal Ministry of Posts and Telecommunications, and the latter, the responsibility of the Deutsche Telekom.

Second, for the telecommunications market, competition is the rule and monopoly the exception, which has to be justified. Therefore, as a general rule, all telecommunications services and all telecommunications terminal equipment are offered in competition. Deutsche Telekom can take part in this competition. In telecommunications, only the network and the telephone service monopolies have been kept to ensure that Deutsche Telekom will be able to perform its infrastructural tasks and maintain its financial strength. Without a certain amount of financial security, it would be difficult to finance telecommunications investment, particularly in the new Federal States in the eastern part of our country.

Third, it is also increasingly possible for private providers to operate in fields that are basically part of the telephone service monopoly. Thus, competition takes place in the fields of mobile communications and satellite communications.

Challenges

Under these general conditions and against the background of rapid technological advances, a worldwide trend toward deregulation, and the increasing significance of telecommunications, Deutsche Telekom currently faces three major challenges:

> to strengthen the competitiveness and expand the range of services,
> to internationalize the services and respond to the globalization of customers, and
> to modernize the telecommunications infrastructure in the eastern part of Germany.

Concerning the first challenge, I have already drawn your attention to the development of global networks and the competitive efforts undertaken by Deutsche Telekom. Because we are going on the assumption that monopoly services, as a percentage of overall turnover, will decline substantially within the next few years, we will have to further strengthen our market orientation, diversification, the provision of new services, the use of the most advanced technologies and the reduction of costs.

As to the second challenge, Deutsche Telekom has established subsidiaries in Japan, the United States, the United Kingdom, France, and Belgium. We do not however, intend to participate directly in projects that deal with the operation of telecommunications networks abroad in the foreseeable future. The exception is, at present, Eastern Europe, where we will also make investments to develop the network. On the other hand, Deutsche Telekom offers to act as a hub for telecommunications links between non-European countries and countries in Eastern Europe. In principle, we are also interested in international projects involving promising new technologies and marketing know-how if there are advantages to be gained from them for our customers. We are, therefore, open to forming pertinent strategic alliances.

Concerning the third challenge — the restructuring of the telecommunications infrastructure in the Eastern part of the Federal Republic of Germany — we are facing a threefold task:

> We have to convert a technology-driven administration into a market-driven enterprise.
> We carried out one of the biggest mergers in recent times when we acquired the telecommunications part of the former German Democratic Republic postal administration; that means 43,000 people. These colleagues had been living and working under completely different conditions until now and we have to make them feel at home.

Finally, we have to replace an infrastructure that is completely outdated. The integration and/or replacement of networks and technical installations includes parts that are between 25 and 60 years old. We must process 1.3 million old applications for telephone connections, some of which were filed 15 years ago, as well as a flood of new applications for voice and nonvoice connections. The first big success was the digital overlay connection between East and West Germany, which became operational on July 7, 1991. Last, there is the need to completely reorganize and integrate telecommunications in Eastern Germany.

In order to cope with these enormous tasks, we have initiated a development program called "Telekom 2000," with more than DM 55 billion earmarked for investment over a period of seven years. When this program has been completed, the telecommunications standard in the new Federal States will be comparable to that of the western part of Germany.

Not everything in this huge effort will be done by Deutsche Telekom. Under the Telekom 2000 program, contracts for turnkey projects will be awarded to private companies. Therefore, the prime contractors will be responsible for all the services required for expansion of the subscriber line network, including switching transmission, building construction, radio relay, power supply, and cabling. These prime contractors will have much scope for action, as rapid implementation will be given top priority. Also, private companies will have the possibility, effective immediately, of providing the monopoly telephone service via satellite networks within the eastern and with the western part of Germany. Private satellite networks may be interconnected with the telephone network of the Deutsche Telekom and the West without special access tariffs being charged.

In following this course of action, Deutsche Telekom can rely on a broad range of vital assets: it has a senior management team with a successful track record in business; it has the know-how and the financial resources to translate technological advances into visible advantages for its customers; it has far-reaching experience with providing sophisticated communications infrastructures. Thus, there is a good chance that Deutsche Telekom will deal successfully with the three challenges, that is, its internal reorganization in terms of strengthening the competitiveness, its investment program in Eastern Germany, and the expansion of its international presence on the basis of cooperative strategies.

4

Recent Developments in UK Telecommunications Policy

Adrian Norman

United Kingdom (UK) telecommunications policy has changed out of all recognition in the 1980s. In 1979, the Tory government under Margaret Thatcher replaced Labour. Thirty-five years of consensus on the merits of a "mixed economy" and collective rather than individual responsibility were swept aside. The most visible consequence was the decline of trades union power, initially in manufacturing, mining, and transport and later in telecommunications, newspaper publishing, and broadcasting.

The most successful changes in telecommunications policy were introduced quickly as an act of political will. The institutional adjustments to cope with the political changes have been less successful.

A simple chronology of al the significant policy changes and administrative actions fills several pages, demonstrating the breadth and complexity of the industry and its ramifications. No one in government in the United Kingdom has an overall understanding of the whole information technology and services area, nor any overall responsibility. Knowledge, like responsibility, is therefore fragmented, and policy develops without full consideration of all the consequences. Not surprisingly, the successful innovations have been those that have reduced government's need for knowledge on which to base interference with market forces; the failures have resulted from the exercise of authority in a limited domain while ignoring the implications in other domains.

Liberalization and privatization of public telecommunications have worked because they were predicated on the entirely political view that governments did not know enough to interfere with market forces. Cable and satellite broadcasting policy has not worked as intended because politicians and officials felt obliged, again for political reasons, to interfere despite their unavoidable lack of understanding of developments in technologies, markets, and institutions.

One peculiarity of UK practice warrants a brief mention. Detailed knowledge of pertinent new technologies, such as cryptography and signal processing, is often classified "secret" because of its defense applications. Companies and officials are inhibited from applying this knowledge in civil markets, and military specialists seldom transfer to civilian work in business or in government. Where government is involved with industry, officials have been much better able to assess cost than value and to perceive need better than opportunity for others to exploit, and policy has worked. Where they have stepped in with restrictive licenses, opportunities have been lost.

This chapter looks at a little more than a decade of policy making for telecommunications in the United Kingdom. It treats the subject broadly because its breadth is the challenge it presents to policy makers. Therefore, telephony and television, satellites and cables, one-to-one and broadcast, switches, and local area networks (LANS), value added networks (VANS), and video cassette recorders (VCRs), and many more topics are included, with customers, subscribers, carriers, and equipment suppliers.

By 1980, the technological distinctions between telecommunications, computing, broadcasting, and publishing were blurred. Telecommunications networks carried digital signals between computers and intelligent equipment on customers' premises. Computer systems were geographically distributed and interconnected by telecommunications networks. Still images were digitized, transmitted, and processed; desktop publishing was imminent, and professional publishers converted to single keystroking, electronic prepress technology and facsimile distribution to remote printing centers, databases, and hypertexts. Digital techniques invaded film and television (TV) studios, initially for animation and later for editing. Broadcasters laid plans to follow publishers from digital preparation to digital distribution of high-definition television (HDTV). Multiple Subnyquist Sampling Encoder (MUSE) in Japan, and multiplexed analogue component (MAC) in the United Kingdom (and later in Europe) left the user confused. Music went digital on compact discs but not on digital audio tapes, a technology so good for copying that the owners of copyrights tried to ban it. Music and video publishers fell prey to equipment manufacturers who recognized the dependence of the media markets on the messages they carried.

By 1990, the stack of hi-fi equipment cabled together to play the content of broadcasts, discs, cassettes, tapes, synthesizers, and other audio carriage media was the model for video media. The color computer screen and HDTV receiver will merge; digital movies will be compressed onto digital VCRs and optical discs; adequate black-and-white TV images will squeeze down 64 kilobits per second Integrated Services Digital Network (ISDN) channels to work stations supporting multimedia communications and processing; broadband satellite and cable networks will carry HDTV quality images in channels a few megahertz wide. Publishers are drawing together media empires to exploit the capacities and capabilities; chip and component makers are tooling up for gigahertz, gigabytes, and giga instructions per second (Gips) in the 1990s just as they planned for megahertz, megabytes, and mega instructions per second (Mips) a decade earlier.

Governments, regulators, and standards makers follow panting in pursuit of converging and colliding technologists and marketers. Perfect copying threatens copyright; distance independence changes the economics of tradeable services, leaving a nation's vital interests vulnerable to the domestic politics of its trading partner; satellites broadcast across sovereign borders, causing national legislators to try unavailingly to repeal the laws of physics; hackers invade foreign computers from a safe haven where telecommunications are a decade ahead of the law; standards are seen to determine the size and profitability of future markets, but the standards process lags behind technology rather than leads it. From all sides, the demand is for government action to protect interests established under earlier regimes and threatened by new developments in foreign countries or converging businesses. The future, with no champion present to make its case, would be the loser if government were able to accede to such demands. The response time of government to fundamental changes in markets is approximately a decade; the response of the market to technological change is almost as long. In 1990, the UK government had adjusted to the technologies of 1970: digital net-works, satellites, teletext, mobile communications, large-scale integration (LSI), databases, digital recording, and the convergence of computing and telecommunications. It has yet to respond to the emerging technologies of 1980 on which are based the markets of the 1990s, such as, expert systems, interconnected global networks, safety critical software, digital audio and video, very-large-scale integration (VLSI), multimedia databases, computer-assisted mis-chief, just-in-time information, civil (low-cost) use of military (high-cost) positioning and cyptography techniques, and the convergence of content and carriage.

The technologies of telecommunications, computing, and broad-casting converged in the 1970s, but policy making in the United Kingdom still retains its links with the past. The result has been a lack of continuity and coherence between and within departments, leading to intervention at cabinet and cabinet-office level at internals. National policy has been made by a very few very busy people on the basis of a few short documents. The views of traditionally strong departments, Her Majesty's Treasury (HMT), Ministry of Defence (MOD), Home Office (HO), and the Department of Trade and Industry (DTI), have made policy decisions. The average term of office of a secretary of state at the DTI has been a year, too brief to master the complexities of telecommunications policy and too unrewarding to fight a difficult corner as sponsor of the industry against MOD and HMT, concerned with competition and efficiency, and against HO, defenders of noncommercial values.

The lack of parliamentary understanding of technical issues, which is compounded by the almost total absence of staff to advise backbench and opposition members, has prevented the growth of an alternative source of informed policy proposals. The United Kingdom has, therefore, followed the United States somewhat blindly, drawing on the products of the latter's long tradition of policy research but lacking a framework in which to view the alien experience.

Industry recognized the opportunity for lobbying presented by the weakness of the policy process in government. Established companies needed to devote only a tiny fraction of their marketing effort to safeguard and enhance their position by lobbying. The return on similar effort in any other part of the business would be negligible. New enterprises could not match the old guard (such as GEC, Plessey, and BBC, and the press barons), particularly when the newcomers' goal was the removal of barriers to entry of start-up businesses. Only in markets with doubtful prospects, such as cable television, have outsiders been welcome.

Faced with a mass of detail beyond its comprehension, government set up regulatory authorities to distance itself from decision making in areas where a market is not effective. The Office of Telecommunications (OFTEL) monitors and enforces licenses under the Telecommunications Act, 1984, safeguards effective competition, and advises on new licenses. There are similar bodies for radio broadcasting, television on all media except videocassette and spectrum allocation, and other natural monopolies such as water, gas, and electricity. While they are young and well led, they can preserve their independence from the industries they regulate and from government because they answer ultimately to Parliament itself, not to ministers. (A government with a solid majority in Parliament can always enforce its will on a recalcitrant regulator but risks paying a high political price.)

In 1979, British Telecom (BT) was starting to introduce System X, an electronic switch, to replace the Stowger exchanges invented in the 1800s. Customer premises equipment (CPE) was supplied by BT: wiring, telephone, fax, small PABXs, telex. Only the biggest customers could shop around among BT-approved suppliers, whose biggest market was BT itself, for large PABXs. Analog technology dominated, except for trunks and junctions. The interface to the past constrained the design and application of new equipment: touch-tone phones generated dialing pulses; area codes differed according to the location of the calling station; itemized billing was impractical.

Independent evaluators rated System X very highly, but it was BT's solution for the UK network, not their suppliers' offering in world markets tailored to local requirements. It reflected the excellence of BT's research into technology and applications and its indifference then to its customers' and suppliers' needs. GEC, Plessey, and STC, the main suppliers to BT, failed to build internationally competitive products based on System X and lost their independence within ten years. STC and BICC, the main cable suppliers, were more successful both on land and undersea. In satellite communications, BT was of little value to BAe as a pioneering customer for orbiting hardware because it relied on Intelsat, Inmarsat, and Eutelsat, but UK companies did win earth station business.

The System X program of exchange replacement was constrained by the external funding limit (EFL) imposed by HMT on BT. In effect, BT's share of national investment funds was insufficient to meet the cost of the exchanges even though the project was very profitable. As a government department, it could not borrow private capital independently of the treasury, since the treasury was its guarantor and competitor in the market for funds. Had BT been privatized a decade earlier, it would have completed the digitalization of the network a decade earlier; it lacked money, not markets or technology.

The network is now essentially digital from the local exchange inward, but twisted copper wire pairs still provide almost all the local links. BT will not carry the broadband network to every home, nor will ISDN be universally available. After a false start in the mid-1980s, ISDN is now spreading. Many businesses have had private networks working to ISDN standards for several years, using equipment from many sources. They have also had LANS interoperating with national and global wide-area networks, experience that has raised the competence and level of ambition of telecommunications managers. The requirement and opportunity to choose equipment and services, not merely to purchase such services as were available after long delay, has forced customers to learn. Educated customers are most demanding and less tolerant of poor performance.

Domestic subscribers have a choice of CPE from around the world at prices as much as an order to a magnitude lower: £300 today will buy a fax better than any available for the equivalent of £3,000 in 1979; cheap telephones can now be bought on the high street for less than BT's quarterly rental then. Chips not only improve performance, they also cut costs.

Radio and TV receivers for terrestrial broadcasts have always been available from retailers independent of the BBC, independent TV (ITV) companies, and other broadcasters. Designs have been in the public domain, and patent protection has not inhibited competition. By 1979, few UK homes lacked a color TV set, and transistor radios were plentiful. In the early 1980s, the Sony Walkman and imitators, VCRs, and teletext spread rapidly. Viewdata, which used TV monitors to display data delivered by telephone, found niche markets only, unlike Minitel in France.

Backward compatibility with existing receivers limited the scope for new technology in broadcasting. Her Majesty's Government (HMG) even had a struggle to phase out 405-line black and white broadcasts to 20-year-old sets when it wanted to release the frequencies for mobile communications. Both satellite and cable distribution, like VCRs, had to serve the installed base of TV displays, not a new generation, despite their higher bandwidths.

Cable distribution in 1980 merely relayed the three terrestrial services to areas where reception was poor. There were hardly any out-of-area signals to import and no established demand for local programming. HMG saw cable TV as an alternative local telecommunications network to compete with BT's copper wire pairs. It imposed on cable carriers technical standards that permitted many other services in addition to TV distribution, then it denied the carriers the licenses to provide telephony except in conjunction with BT or Mercury Corporation Limited (MCL), the duopolists. The duopoly policy, which applied to a large, established industry, won the Whitehall battle against policy for the small new industry.

This pattern of promoting new developments and then failing to support them effectively is well illustrated by the story of direct-to-home satellite TV. In the early 1980s, a new signaling system was developed with MAC, which would support better TV pictures using the higher bandwidths available on satellites and cables. One of the conditions of the Business Satellite Broadcast's (BSB) license from the IBA in 1986 to provide the UK direct broadcast satellite (DBS) TV service was the use of B-MAC, a condition later endorsed by the CEC, who required MAC for all DBS in Europe to pave the way for HDTV in the mid-1990s. BSB had to develop a receiver that would translate B-MAC to the TV broadcast standard used in the United Kingdom (PAL) so that viewers could use their existing displays. The same receiver handled pay-per-view and pay channels, with over-air addressing and enabling. MAC also carries digital sound and data, opening opportunities for other telecommunications services to business and homes.

HMG, having established (on DTI's advice) technical standards in cable and satellite broadcasting that opened the way for new telecommunications services on the back of entertainment, then let other considerations frustrate their foresight. Sky Television was launched by the newspaper proprietor Rupert Murdoch after BSB had won the franchise for direct broadcast satellite TV in the United Kingdom. It used the Luxembourg satellite, Astra, which was not subject to regulations about MAC governing DBSs, to transmit a signal on telecommunications frequencies at sufficient power for 60-centimeter domestic antennas to pick it up. Sky broadcast in PAL, so no translation was needed from MAC. It had no commitment to the IBA to maintain quality, therefore, it used cheaper material; and it was not pulling through new services, standards, and technology for the next generation. Not surprisingly, it was quickly to market, launching its service while BSB was still trying to make its new technology work. HMG, which had protected Mercury for seven years against competitors other than BT while it got established, did not protect BSB, to whom the IBA thought it had given a monopoly of the UK direct-to-home (DTH) market. In 1991, Sky and BSB merged before each destroyed the other, but the new standard has been a casualty.

The future of MAC and of HDTV in Europe is under intense discussion at the Commission of the European Community (CEC) as this is written. The problem of European technical standards is well understood and much researched. The debacle of PAL and SECAM, the rival standard adopted for the last generation of TV by the United Kingdom and France and their respective followers, provides an object lesson. A single European standard is essential if the market is to be large enough for 150 million receivers to be made in Europe (not imported from the Far East at a cost of £100 billion over ten years). However, the adoption of one standard requires banning of others, an interference with the market that runs counter to accepted policy.

To add to the confusion, fully digital HDTV is now on the horizon for 1998, spelling the end of the multiplexing of analog components, the limit of technical capability in 1983. The CEC's decision in 1986 was right, but it was not accepted by policy makers.

THE MAIN SEGMENTS OF THE
COMMUNICATIONS MARKET

In 1977, HMG received reports from commissions on three separate areas of policy that had raised issues too difficult for ministers to resolve. Carter reported on the post office, which at that time still ran BT. McGregor looked at the press, concentrated in Fleet Street and tied by its unions to archaic technology and practices. Annan dealt with broadcasting, then restricted to the BBC's two channels and the ITV's single channel divided into regional services. In the same year, a World Administrative Radio Conference

(WARC) addressed DBS, microchips became a matter of public debate, and the first personal computers appeared. The merger of the three separate market segments was now inevitable.

The BBC had been launched more than half a century earlier with a charter "to inform, educate and entertain," early recognition of the multipurpose capability of carrier technology. The gramophone record was the only storage medium for sound until the "talkies" added sound to film. Live broadcasting from studios was the norm for radio and TV until the 1950s, except for films and records. Storage at the receiving end for time-shifting broadcasts is a recent development, influenced as much by audio and video prerecording as by broadcasting. With the arrival of digital technology, chronocommunications (across time) and telecommunications (across distance) will share equipment in multipurpose systems designed primarily for entertainment but able to support information and education also.

Publishing systems, including those used by the press, now use similar digital technology to achieve chronocommunication and telecommunication for the same three markets: information, education, and entertainment. They use paper as the primary storage and display medium, whereas broadcasting uses screens, tapes, and discs.

The post office, investigated by Carter, carried paper in the mails and telecommunications over wire and wireless links. He recommended that the different technologies compete, but it was not until the Tories supplanted Labour that HMG split BT from the post office and set in train a long series of reforms in telecommunications and broadcasting. Meanwhile, the press reformed itself and pushed into broadcasting.

TELECOMMUNICATIONS CARRIAGE SERVICES

BT now, as then, focuses essentially on the carriage of information without concern about its content. The key decisions in fixed and mobile point-to-point terrestrial telecommunications are:

1980–81 — Liberalization
  British Telecommunications Act separates BT from post office
  HMG sells 49 percent of Cable and Wireless (C&W) shares (and rest in 1983 and 1985)
1982–83 — Privatization
  MCL licensed to run network in competition with BT
  British Approvals Board for Telecommunications (BABT) established and CPE liberalized
  BT Securicor licensed to offer cellular radio
  General license to supply VANS on UK public networks

HMG gives seven years' protection to duopoly from November
  1983 to allow MCL to get established and BT to adjust to
  competition
Duopolists cannot provide TV on their main networks but
  may join in other consortia, and cable TV operators cannot provide voice services without BT's or MCL's participation
1984 — Reregulation
The regulated duopoly created by the Telecommunications Act
  requires licenses to run telecommunications systems
Private Telecommunications Operators (PTOs) have rights and
  obligations
Other operators hold class or individual licenses, for example,
  for value added data services (VADS), and branch system
  general license (BSGL)
No simple resale of leased lines until July 1989 at earliest
Sale of 51 percent of BT in December
1985 —
Installation and maintenance of first residential telephone
  opened to competition
Interconnection determination gives Mercury right to connect
  to BT network
BT acquisition of Mitel takes it into equipment manufacturing
1986 — Mobile services
Licenses for two nationwide private mobile radio (PMR)
  networks, several regional networks, and two national
  paging network operators
October: United Kingdom now largest cellular operator in
  Europe, with 100,000 customers
1988 — BT prices pegged to 4.5 percent below inflation
1990 — Duopoly review and government response
1991 — Entry of many new competitors

TELECOMMUNICATIONS CONTENT SERVICES

The key decisions in broadcasting are:

1977 — WARC and Annan Report
1980–81 —
Trades union power curbed, opening way for new technology
  and working practices in the media
Teletext promoted by joint action of government, broadcasters,
  set makers, and retailers
HO publishes white paper on DBS
1982 —
Channel 4, ITV second channel, starts up

1982 Information Technology Advisory Pannel (ITAP) on Cable
  Systems recognized interdependence
Hunt report
1983 — HO and DTI joint report on broadcasting policy (April);
  Broadcasting Bill establishes Cable Authority, allows for DBS;
  Unisat project UK DBS begins under BBC leadership
1984 — Budget ends capital allowances to cable companies
1985 —
Cable Authority starts
1985–90: slow growth of cable TV; consumers find it no better
  than terrestrial TV from national BBC and ITV services;
  VCRs are cheaper and offer more choice
1986 —
Peacock Report on financing the BBC
Murdoch revolutionizes Fleet Street, moving his newspapers to
  Wapping, sacking the printers, and introducing single
  keystroking
December: BSB gets DBS franchise
1988 —
Murdoch launches Sky TV on Astra telecommunications
  satellite from Luxembourg direct to homes in the United
  Kingdom
Broadcasting White paper; DTI presses for competition, choice
  and industrial efficiency, while HO defends quality and
  culture
1990 — Broadcasting Act establishes new regime for ITV (not BBC),
  independent radio, cable, teletext, satellite
1992 — New International Telecommunications Companies (ITC)
  franchisees take over
1996 — December: BBC charter runs out

Omitted from the above is any consideration of radio regulation
and spectrum policy. The responsibility was passed from HO to DTI
in the mid-1980s. Enquiries and reports contributed to its policies,
which crucially affect broadcasting, mobile radio, satellite, and local
distribution services.

TELETEXT

The use of broadcasting media for telecommunications carriage
services does not fit conveniently with the above chronology, because
teletext on the vertical blanking interval between TV frames was
devised in the early 1970s and standardized so that any broadcaster
could send text for display on any suitably equipped set. Both the BBC
and ITV companies established services. Public teletext now provides
some 1,500 pages broadcast about ten times per minute with TV

listings, weather, sports results, advertisements, and similar material. In 1980, less than 1 million sets could receive teletext; now there are 10 million equipped to do so.

Because the data displayed on the screen were transmitted as binary digits, they could just as easily carry information for private computers as for public display. Indeed, teletext was used for software distribution to schools and homes for some years. As a broadcast medium, however, all material had to be for general reception. This has not stopped the BBC and ITV companies from offering "private" teletext, the customers relying on the messages being meaningless to anyone who chooses to intercept them. Retailers use this data broadcasting service to distribute warnings of hot charge cards, shelf price changes, and administrative notices. Financial data, horse racing betting odds and results, and travel industry notices are also carried.

In 1990, the Broadcasting Act allowed the broadcasters to carry telecommunications services on broadcasting frequencies; therefore, they can now offer encrypted, and thus truly private, messaging, subject to licensing under the Telecommunications Act. This opens the possibility of private broadcasting of messages to homes, once the preserve of pubic broadcasters. The distinction maintained so far between carriage and content services in UK legislation is blurring, but it was still clear in the 1980s.

THE MAJOR SUPPLIERS OF EQUIPMENT AND SERVICES

In 1979, each industry had a few dominant players. BT had the exclusive privilege to provide all public telecommunications, with the exception of paging. It represented the United Kingdom on Intelsat, Eutelsat, and Inmarsat and attended international conferences as a government department. Three companies, protected from international competition as national champions, supplied almost all its switching equipment; GEC, Plessey, and STC supplied almost all its needs. Transmission was no more competitive. There was little competition in CPE. C&W operated outside the United Kingdom only and was also partly government owned.

At the same time, ICL was the national champion in computing and had a privileged position as supplier to government and to public administration and nationalized industries who could be persuaded to "buy British."

In broadcasting, the BBC and the ITV companies had their respective monopolies of license fees and advertising on TV.

In 1991, all the incumbents from 1979 faced competition. In telecommunications, there are newcomers in all fields.

In the backbone network, MCL's seven years' protection has ended, and several companies with networks and "code powers" (to lay ducts for other services across public and private land) in other industries have declared their interest in competing: British Rail; London Transport with France Cable Radio; the water, gas, and electricity utility network operators; and National Transcommunication Limited (the independent TV network operator).

Seven specialized satellite services operators offer UK and European services in satellite news gathering, business television, data broadcasting and messaging, video conferencing, and perhaps other new services: BAeCom, SIS, EDS, MSC, BSB DataVision, Uplink, and Kingston Communications. Reuters has an interest in Uplink, France Cable Radio in Maxwell satellite communications, and BAeCOM in several other new telecommunications businesses in addition to satellites, which it builds.

In mobile communications, Racal Vodaphone and CELLNET are well-established cellular telephone operators with the largest subscriber base in Europe; there are several PMR companies and four embryonic public mobile data services; telepoint services have started, and personal communication networks are imminent; paging services have proliferated; a second generation of cordless telephones is coming in, and the pan-European Groupe Speciale Mobile (GSM) and DECT standards are in place, therefore, international roaming will be possible.

The personal communications network consortia show the international interest in the UK market: Mercury PCN Ltd. is owned by C&W (MCL's parent), Motorola, and Telefonica; BAeCOM has joined with Matra, Millicom UK, Pacific Telesis, and Sony; and Unitel Ltd. combines the resources of STC, Thorn EMI, US West, and Deutsche Bundespost Telekom.

Wireless private branch exchanges (PBXs) and cordless phones are available from High Street stores and numerous small installers, who provide services to customers once obliged to rent building wiring and CPE from BT. Cheap handsets, approved by an independent board, cost less to own than BT once charged for a year's rental.

In broadcasting, the terrestrial network faces reorganization and competition from satellite, cable, and video cassettes. A fourth terrestrial channel, operated nationally by the regional independent TV companies, started in 1981, and a fifth, serving most but not all of the country, is scheduled to open soon. BSkyB has combined British Satellite Broadcasting's five channels with Sky Television's several channels on Astra, the Luxembourg "HotBird," which also carries other channels in English and continental languages. European Space Agency's (ESA) Olympus carries experimental and educational programs directly to homes, and other satellite "broadcasts" can be picked up with amateur dishes by interception of signals intended for cable headends.

Cable franchise areas advertised by the Cable Authority (CA) were well into three figures before the CA was merged at the end of 1990 with the ITC.

Significantly missing from the list of competitors in 1991 are any large information technology (IT) companies that had grown from small beginnings since 1979. MCL is owned by C&W and BAeCOM by BAe. BSkyB belongs to media giants. RACAL Telecommunications grew within a successful large IT business, RACAL Electronics.

THE DRIVING FORCES FOR CHANGE

For two decades, public policy defied the forces that doomed it to fundamental change: technology, market, and foreign experience. Technology took no account of national borders, but a monopolist need not use what was not invented here; market demand could not be suppressed, but supply need not rise to match it when competition can be suppressed; and foreign influences may be safely ignored if foreigners, like natives, cannot establish businesses to promote them.

Technology

In the late 1950s, digital transmission and switching were firmly established as the inevitable successors to analog transmission and Strowger switching in the public telecommunications network. At the same time, telecommunications were tying together computers in networks. Optical fibers, satellites, microwaves, integrated circuits, signal processing, software engineering, cryptography, and other ITs invaded the network in the 1960s and 1970s. Users started to exercise choice between public and private networks with their intelligent systems, first for data and later for voice services. Software in the CPE helped users get round the monopolist's restrictive practices.

At the same time, de-skilling and disintermediation made networks and systems available to end users directly. Single keystroking followed subscriber trunk dialing, networking in offices eliminated messengers; a work station on every desk waited for its user, doing away with queues for access to big computers; computers in homes and schools removed the mystery surrounding the specialist; and "chips with everything" took IT into household appliances.

The "I" of IT was affected by the "T." Consumer electronics changed the TV and music industries. Publishing moved to desktops; newspapers moved to buildings without human printers and their unions. Secondary and electronic publishing are now significant industries, often entwined with messaging services like Reuter's Monitor and Prestel (BT's less successful version of France Telecom's [FT] Minitel).

Markets

Choice, already familiar to consumers from the computer and publishing worlds, spread to telecommunications and broadcasting in the 1980s. Consumer electronics had always offered choice: the music industry had records under many labels; radios, TVs, VCRs, satellite dishes, home computers, and other equipment were available from many sources. However, in 1979, telephones, facsimile transceivers, modems, and their attendant services were a monopoly. For every new market in telecommunications, HMG has licensed competitors.

In broadcasting, independent producers and foreign competition now vie with the in-house production that dominated the BBC and ITV channels in 1980. The VCR and video store offered choice and flourished; cable networks relayed network TV and languished despite their local monopolies. However, DBS grew from nowhere to 7 percent penetration in two years and is growing at 100 percent per annum; it offers more choice than the BBC and ITV companies combined. Much of the material is foreign, particularly U.S. and Australian, and is aimed at a wide, unsophisticated market with which regulators, unlike advertisers, cannot identify. The government was forced to introduce a "quality threshold" that applicants for ITC franchises in the 1993 round must meet, despite its original intention to let them to the highest bidder.

Influences from Abroad

The United Kingdom has always been a trading partner in Europe; the United Kingdom has also been increasingly subject to the shared decisions of the CEC during the 1980s. This influence will grow after 1992. In research and development, the European programs are important sources of ideas and test beds for future commercial products and services. Academic and commercial researchers have learned to work together internationally, and technology transfer across national borders has increased. Some of the transnational mergers between European companies have been helped by previous experience of joint projects.

The CEC's Green Paper on Telecommunications in the middle of the 1980s and its subsequent directives shook up the complacent Post, Telephone and Telegraph (PTT) authorities in most member countries and spread the British policy on liberalization. Telecommunications profits no longer subsidize postal losses; provision and regulation are separated; VANs and CPE are available from many competitors; standards and approvals are international; large procurements from only national champions are difficult.

The United Kingdom now looks to Brussels for telecommunications standards, precompetitive research in the Research in Advanced Communications Technologies in Europe (RACE) and other programs, the opening of telecommunications markets to competition after 1992, and the negotiation in the General Agreement on Tariffs and Trade (GATT) of trade issues in this, as in every, area.

The United States

Telecommunications policy in the United States profoundly affected UK thinking at the start of the Thatcher era. The cost of monopoly was made clear by the many U.S. firms based in the United Kingdom, which had always welcomed inward investment and sought more, particularly in financial services. Unlike the United States, there are no significant restrictions on foreign ownership, no equivalent of the Buy American Act, and no extraterritorial application of domestic law As a result, many U.S. companies have participated in the development of UK telecommunications and broadcasting, particularly in cable networks, in which they perceive opportunities that are still hidden from domestic investors. UK companies are finding the U.S. market harder to crack, but BT and C&W, with their global ambitions cannot ignore it. The former has bought in; the latter merely interconnects its network.

Japan

The United Kingdom is as welcoming to Japanese investment as it is to U.S. or European, which distresses some of its European Economic Community (EEC) partners. HMG and the British public feel that it is better to meet some of the insatiable demand for Japanese consumer and business electronics (and cars) from local factories, a precedent set by IBM and Ford among many others. Shares in British companies can be traded freely. Fujitsu has acquired ICL Private Limited Company, Mitsubishi owns Apricot Computers, and other foreigners own parts or all of many UK businesses.

The Japanese fifth generation computer program inspired the United Kingdom's Alvey program of national collaborative research into new IT from 1983 to 1988. The impressive result were as quickly absorbed by foreign observers as by British participants; therefore, the UK effort now goes into European collaboration rather than into purely national efforts.

The United Nations, Its Agencies, and Other International Bodies

The United Kingdom participates fully in the International Telecommunications Union (ITU), International Organization for Standardization (ISO), and other international bodies involved in telecommunications. It was never a member of the Intergovernmental Bureau for Informatics (IBI) and withdrew from the United Nations Educational, Scientific, and Cultural Organization (UNESCO) some years ago. BT no longer represents the United Kingdom at governmental level, unlike its predecessor body. It has, however, retained its role as signatory of the Intelsat, Eutelsat, and Inmarsat treaties; a separate Signatory Affairs Office in BT now handles relationships with these organizations impartially on behalf of BT and its competitors.

The UK position on the complex policy issues in international telecommunications are outside the scope of this chapter. Suffice it to say that BT seeks to be a national champion in global markets, in which C&W is already an established player. HMG wants to spread the gospel of competition and liberalization, particularly the opening of foreign markets to UK companies. In the GATT, in which the CEC represents the United Kingdom, this position is not shared by all its European partners.

REFERENCE

Palmer, M., & Tunstall, J. (1990). *Liberating communications: Policy-making in France and Britain*. London: NCC Blackwell.

5

Impacts of the 1985 Reform of Japan's Telecommunications Industry on NTT

Hajime Oniki

Japan's telecommunications industry was substantially reformed in 1985. Before that time, Nippon Telegraph and Telephone Corporation (NTT) was the state-owned common carrier operating as monopoly for domestic telecommunication. Kokusai Denden Corporation (KDD), (that is, the International Telephone Corporation), having been privatized in 1952, was the monopoly common carrier for international telecommunication. In April 1985, competition was introduced by privatizing NTT and allowing three new common carriers (NCCs) to operate nationwide for long-distance telephony and other new carriers to operate regionally or with mobile telephones. Local telephone markets, however, were left under NTT's monopoly. International telecommunication in Japan became competitive in October 1989, when two NCCs were allowed to enter the international telephone market.

The 1985 Telecommunications Business Law of Japan recognizes two categories of carriers: type I and type II. They are distinguished by their facilities, not by their services. Type I carriers are those operating with physical transmission circuits; type II carriers are those without circuits. In effect, a type I carrier can offer every service a type II carrier can, but not vice versa.

Type I carriers are regulated by the Ministry of Posts and Telecommunications (MPT) in entry-exit, service provision, pricing, and other aspects. Foreign owners are allowed to obtain, in total, one-third of the shares of a type I carrier. In 1990, there were 59 domestic type I carriers, including NTT, and 3 international type I carriers, including KDD. NTT and KDD are established by the NTT Organization Law and the KDD Organization Law, respectively;

they are regulated by MPT more heavily than are other type I carriers.

Telecommunications business by type II carriers was liberalized in April 1985; in particular, telecommunications service trade by type II carriers, import or export became free. No restriction is imposed on foreign ownership of a type II carrier. There is free entry to the industry, and there is no formal restriction on pricing or operations including straight resale of leased circuits by type II carriers.

The Telecommunications Business Law defines two subcategories of type II carriers: general type II carriers and special type II carriers. Special type II carriers are those operating internationally or those operating with 500 or more circuits (measured in terms of the unit equivalent to the capacity of 1,200 bits/second). There is a slight difference in regulation of general type II carriers and special type II carriers in entry into the industry. General type II carriers need only to report of their entry to MPT; special type II carriers need to register themselves with MPT. MPT may reject application for registration by a special type II carrier when the applicant is judged not qualified according to a set of criteria stated in the Law. All in all, however, this difference is not significant, and we can consider that business by type II carriers is liberalized in Japan, at least formally.[1] Since 1985, more than 800 type II carriers have been established, including 28 special type II carriers, of which 16 are operating internationally.

MPT'S POLICY SINCE THE 1985 REFORM

MPT's regulation of Japan's telecommunications industry since the 1985 reform may be summarized in two statements: to facilitate competition in long-distance and mobile telephony, MPT allowed NCCs to operate in an environment more favorable than that in which NTT and KDD operate; MPT encouraged NTT to improve its internal efficiency and to accelerate "digitization" of its network.

MPT, having realized that effective competition in the telecommunications industry is difficult to achieve because of the monopoly power of NTT, especially in local markets, has indicated the possibility of dividing NTT into a small number of carriers.[2] MPT's plan to divide NTT, however, did not materialize in 1990, when a report was issued by MPT to review the present state of NTT and to propose future plans regarding reorganization of NTT[3] (see Figure 5.1). In March 1990, MPT ordered NTT to improve business and technical practices that had prevented NCCs from competing with NTT on equal ground; further, MPT ordered NTT to divest its mobile telephone operations (including automobile telephones, marine telephones, and wireless paging services) in two years.

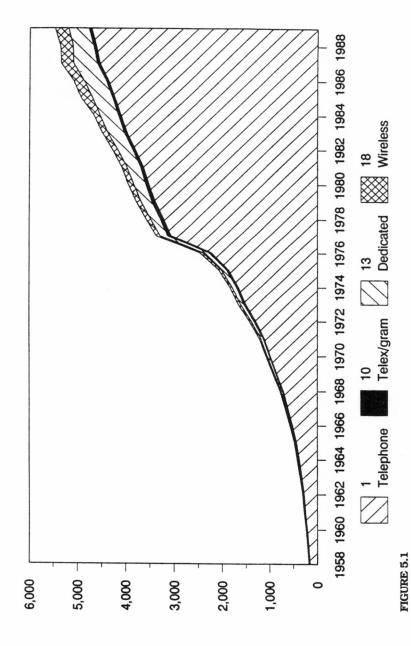

FIGURE 5.1
Operating Revenues (NTT) (per billion per year)

1	10	13	18
Telephone	Telex/gram	Dedicated	Wireless

Since 1985, MPT has maintained a principle to allow a small number of type I carriers to operate in each of Japan's telecommunications markets. In the domestic long-distance market, three NCCs were allowed to enter, and in the international market, two NCCs were allowed. In the regional telephone markets, at most one operator was allowed to enter to compete with the incumbent carrier, NTT. In 1990, there were seven regional type I carriers, among which only one carrier, Tokyo Telecommunications Network Corporation (TTNet), a subsidiary of the powerful Tokyo Electric Corporation, offers public services as well as leased circuit services; all other regional carriers offer leased circuit services only. Two carriers were recently allowed to enter into the satellite communications market. For mobile telephony, MPT allowed one new carrier to enter into each regional market to form a market of two operators, the other one being NTT.[4] Thus, MPT's policy on telecommunications carriership is to create regulated markets, each with a very small number of operators. Whether this policy will foster effective competition or whether this policy will work as a means to "protect" incumbent and new operators is yet to be seen.

MPT's policy to assist NCCs to compete with NTT is best characterized by its price regulation. The Law states that the price of services of type I carriers must be approved by MPT. Since 1985, the price of long-distance transmission provided by the three NCCs has been set, on average, at a level lower by 10 percent to 20 percent than the price of the same service provided by NTT. (In October 1989, when international NCCs were started, MPT determined that the price of services given by the NCCs must be set 20 percent lower than that given by KDD.) As a consequence of this policy, the revenue share of the three NCCs in 1990 rose to approximately 3.8 percent in the entire domestic market and to 40 percent in the long-distance market of Japan's business district, the Tokaido corridor, which extends from the Tokyo metropolitan area to the Osaka-Kobe metropolitan areas. Among NCCs, Daini-Denden, which is the spearhead of Japan's manufacturing and finance community seeking to enter into he telecommunications market, has the largest share. In 1990, two of the three long-distance NCCs started to pay dividends to their stockholders.

In addition to regulations on entry and pricing, MPT made a number of attempts to assist NCCs for rapid growth of their business. Some of those attempts include the following: MPT "informally ordered" NTT to let some of NTT's system engineers leave NTT temporarily and work for NCCs during the initial construction period of their long-distance network, as NCCs had few engineers capable of starting a telephone network. Also, MPT, together with other departments of the government, including the Ministry of Transportation and the Ministry of Construction, administratively helped NCCs construct long-distance circuits. For instance, Daini-Denden was able to obtain permissions to build facilities needed for microwave transmission much faster than in other cases. The other new long-distance carriers quickly obtained from local governments hundreds of permissions needed to construct optical fiber lines across public roads, rivers, railroads, and so on.

Although MPT assisted NCCs to enter and operate in the telecommunications market, it did not directly constrain the activities of NTT. Rather, MPT approved most of the proposals submitted by NTT for price reductions, initialization of new services, and investment. For a short time after the liberalization, MPT considered NTT to be the main conductor of modernization of Japan's telecommunications network and encouraged NTT to promote digitization of its system as quickly as possible. In particular, NTT was asked to accelerate replacement of old nondigital switching machines with new digital ones; old machines still usable and in the process of depreciation were replaced with new machines. Consequently, the proportion of digital switching machines to the total number of switching machines in NTT reached approximately 25 percent in 1989 in terms of the number of user terminals accommodated. NTT also was asked to start Integrated Services Digital Network (ISDN) services as quickly as possible; the basic ISDN service (two circuits of 64 kilobits plus one circuit of 16 kilobits) was launched in April 1988, and a higher-grade ISDN service with greater capacity was started in 1989. In 1990, there were approximately 10,000 subscribers to ISDN services, and the number is rapidly increasing. Finally, MPT allowed NTT to establish a large number of subsidiaries in different areas. The scope of businesses covered by the subsidiaries extends from utilizing information included in NTT's yellow books to manufacturing and marketing prepaid telephone cards.

MPT's policy on the activities and the organization of NTT, however, was resteered in 1989. MPT, shortly after the 1985 reform, realized that in all of the telecommunications markets except the paging market, NTT remained as a de facto monopoly, acquiring a 90 percent share or more. NTT was a monopoly in local markets. Furthermore, as NCCs kept complaining, NTT continued to possess technological and marketing advantages in competing with NCCs. For example, a large number of NTT's switching machines are still of the old crossbar type; a customer whose terminal is connected to an old-type machine cannot subscribe to NCC (NCC could not bill the customer for calls placed through an old-type machine). Further, even though a customer subscribes to NCC, equal access is not realized; a customer using NCC for a long-distance call needs to dial four more digits than when using NTT. An example of other complaints filed by NCCs is that NTT obtains information about customers subscribing to NCCs because the circuit connected to such customers must be relayed by NTT, and NTT could use the information for marketing and other purposes, causing unfair competition between NTT and NCCs. From these observations, MPT seems to have reached the conclusion that without changing the present structure of the telephone market, effective competition may never be realized.

The Japanese Telecommunications Business Law, set forth in 1985, stated that a revision of the Law could be called for three years after 1985. In 1988, MPT decided to postpone revision of the Law because it was only two years after the three long-distance NCCs started their operation. On the other hand, the 1985 NTT Organization Law states that a revision of NTT's organization may be called for five years after the Law was implemented. In 1988, MPT asked the Committee for Telecommunications to consider the possibility of restructuring NTT.

In October 1989, the Committee filed with MPT an interim report on Japan's telecommunications industry.[5] This report suggests the possibility of dividing NTT. Three plans are stated: NTT be divided into two entities, one long-distance carrier and one local operator; NTT be divided into one long-distance carrier and several local operators; or NTT be divided into several regional operators, each responsible for long-distance and local operations. In addition, other suggestions are included in the interim report regarding fair competition between NTT and NCCs, internal reorganization of NTT for furthering efficiency, and investment by NTT for accelerating digitization.

The final recommendation by the Committee on restructuring NTT was filed with MPT in March 1990.[6] MPT compromised by agreeing that it was still too early to make a decision on breaking up NTT and that more time would be needed for further investigation and discussion. This decision was affected by political elements. The Ministry of Finance (MOF) opposed breaking up NTT because the price of NTT's stocks would fall (it actually did when the possibility of dividing NTT was reported) and the revenue to the government from selling them would be decreased. The owners of NTT's stocks were opposed for the same reason. Furthermore, a majority of the Japanese business community (represented by Keidanren and other groups) expressed a view that division of NTT would be too radical a solution for promoting competition in the Japanese telecommunications market. The Socialist and the other opposition parties, holding a majority in the Upper House of the Japanese Diet, were considered to be a factor that would stop any attempt to break up NTT because of a close relationship between the Socialist Party and the influential labor union of NTT. Thus, MPT gave up the idea to divide NTT in 1990; it was stated by the government that a further assessment about reorganization of NTT would be done by 1997.

In the spring of 1990, NTT started an advisory committee to consider promotion of fair competition and other matters related to it. A plan was announced in the spring of 1991 that NTT's mobile telephone operations would be divested in summer 1992. (NTT had divested its enhanced service department in 1988; it became NTT Data Communications, Inc., a special type II carrier.) A number of policies for fair competition with NCCs were carried out by NTT; they included additional installation of points of interfaces (POIs), connection points between NTT's network and NCCs' network, regulation of internal use of information on customers that NTT acquired by relaying them to NCCs, improvement in NTT's accounting system so that the revenue and the cost of services would be stated separately, and opening information of NTT's network and customers to the general pubic.

Because of the price difference between NTT and NCCs, the share of NCCs in the domestic long-distance market increased rapidly, especially for calls between the three major metropolitan areas of Japan — the Tokyo area, the Nagoya area, and the Osaka-Kobe area. In 1990, in fact, the share of the three NCCs in the intermetropolitan calls exceeded that of NTT. NTT, observing the rapid decline of its share in the intermetropolitan calls, started considering "rate rebalancing" in Japan. It is generally agreed that a large amount of subsidy goes from long-distance operations to local operations of NTT, although recent statistics show a continuous decline of the subsidy. In spring 1991, NTT announced a plan for an increase in the local rate accompanied by reorganization of message areas. This would have a strong, adverse effect on the operation of NCCs through an increase in the "access charge." It is remarkable that NTT avoided further price reductions in the long-distance market to compete with NCCs (which was what happened in the international telecommunications market) but sought to use rate rebalancing tactics. NTT started charging for directory assistance services in April 1991 to eliminate a part of its cross-subsidization from the long-distance operation. Whether the rate rebalancing planned by NTT will be approved by MPT is yet to be seen.

PRICE REDUCTION BY NTT SINCE 1985

One of the most important outcomes of the 1985 liberalization of Japan's telecommunications industry is a series of price reductions carried out by NTT and KDD. It can be regarded as a textbook case in which monopoly prices are decreased because of the competition from new entrants.

During the period from 1985 to 1991, NTT, with the approval of MPT, reduced the price of four types of services: long-distance calls, leased circuits, mobile telephones, and paging. NCCs started to offer

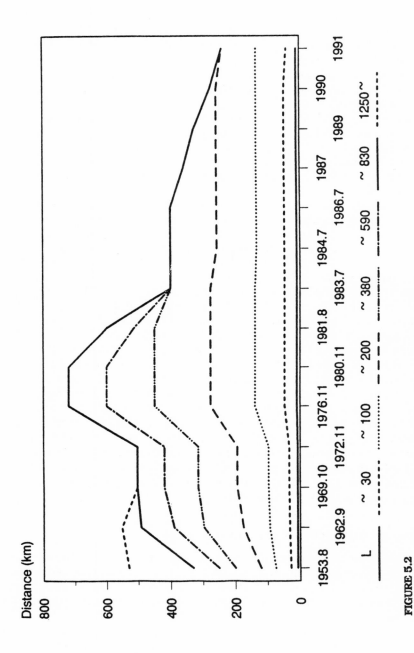

FIGURE 5.2
Telephone Charges — Weekdays Nighttime (NTT: 3-minute charges)

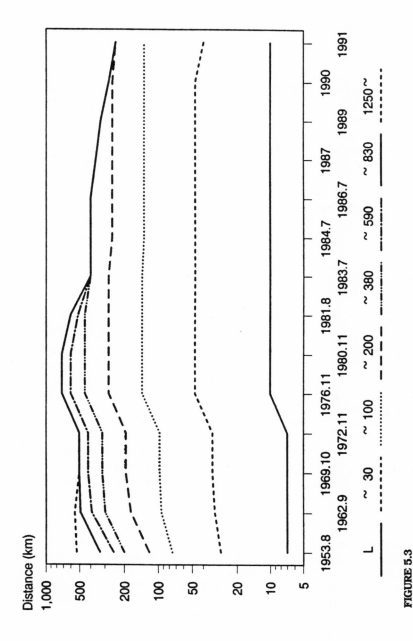

Distance (km)

FIGURE 5.3
Telephone Charges — Weekdays Daytime (NTT: 3-minute charges)

these services in 1987. Figures 5.2 and 5.3 illustrate NTT's telephone charges from 1953 to 1991. It is seen that the long-distance rates were lowered continuously and significantly after the middle of the 1970s. The 1985 liberalization accelerated this trend. Figures 5.4 through 5.7 illustrate the reduction of prices by NTT and NCCs for the four services.

The pattern of price reductions shows the way in which competition was introduced. One can observe a pattern of price reductions common to long-distance calls, leased circuits, and paging. For each of these services, NCCs, at the time of entry, set a price lower by 10 percent to 20 percent than the price being charged by NTT. Soon after the entry, some of NTT's customers switched to NCCs because of the price difference; the share of NCCs increased gradually. Although the share of NCCs was still low, NTT was threatened by the trend in which it lost the share. NTT responded by lowering its prices to regain its competing power against NCCs. Once NTT was allowed to reduce its prices, NCCs felt that they might not be acquiring customers as fast as they did before, and they introduced further price reductions. Such "undercutting cycles" (that is, a price war) were repeated, one after another. From Figures 5.2, 5.3, and 5.5, it is seen that, from 1985 to 1991, in the market of long-distance calls, four cycles took place, and in the markets of leased circuits and paging, three took place.

The pattern in the market of mobile telephones is slightly different. NTT reduced the price drastically soon after the 1985 reform. NCCs started to enter into the market three and one-half years after that; NTT responded with a further price reduction. This is explained by technological progress in mobile telephony (NTT's price reduction in July 1985) and the introduction of competition (its price reductions in December 1988 and March 1991).

In summary, in the market of long-distance calls, NTT's price decreased by 40 percent from April 1985 to March 1991. For leased circuits NTT's price went down by approximately 16.2 percent from April 1985 to May 1989. For mobile telephone services, the price decreased by 57 percent from April 1985 to March 1991. Finally, for paging services, NTT's price decreased by 31 percent from April 1985 to March 1991.

KDD also reduced its service prices during the period from 1979 to 1990; the cost of price reductions was high after 1985, at which time MPT made it clear that competition would be introduced. The largest reduction was made in 1988, one year before two competitors entered the market. Competition in the international market was keener than in the domestic market, because users in the international market are mostly business firms, more sensitive to the price difference between KDD and NCCs than are residential users.

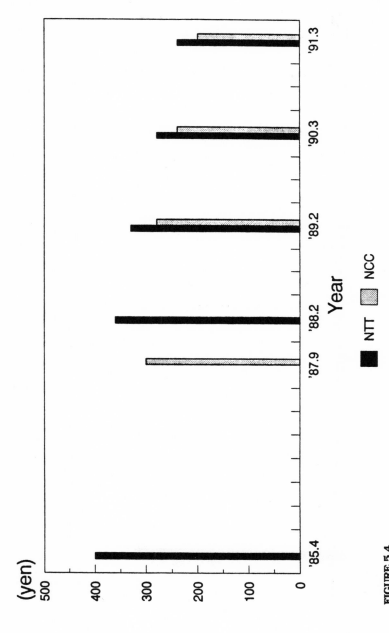

(yen)

FIGURE 5.4
Regular Price of 3-minute Call with NTT and NCC: Tokyo-Osaka

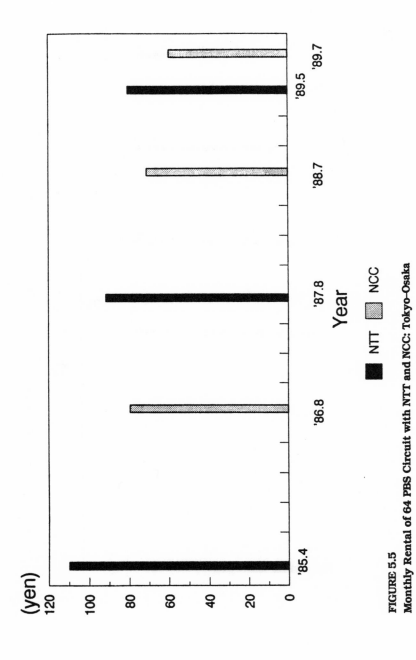

FIGURE 5.5
Monthly Rental of 64 PBS Circuit with NTT and NCC: Tokyo-Osaka

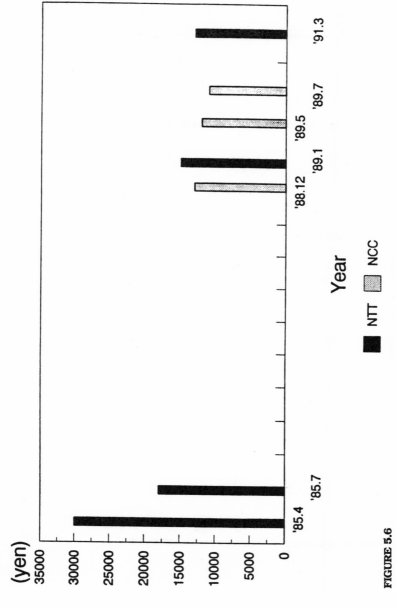

FIGURE 5.6
Monthly Fixed Charge for a Mobile Telephone with NTT and NCC

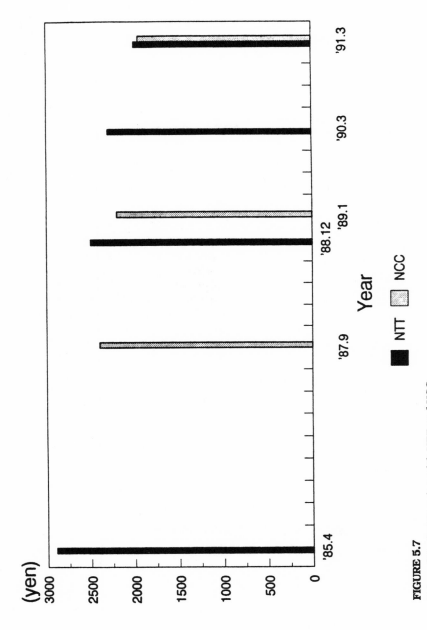

FIGURE 5.7

Price of Paging Service with NTT and NCC

It is clear that the price reductions for long-distance calls were beneficial to telephone users, individual or business. They were made possible partly by technological progress in long-distance transmission and partly by the introduction of competition. In the international telecommunications market, price reductions were so rapid and significant that the price charged by KDD and that charged by NCC became almost equal one year after the introduction of competition. In the domestic long-distance market, however, the price charged by NTT is still higher than that charged by NCC, and the share of NCCs is still increasing. As stated in the preceding section, this alarmed NTT strongly; NTT started considering a number of actions to take to prevent such a rapid change in the share of the long-distance market.

It is now MPT's turn to do something. Roughly speaking, MPT has three alternatives:

To maintain the 10 percent to 20 percent difference between the price charged by NTT and that charged by NCCs without approving NTT's proposal for rate rebalancing; the share of NCCs in the long-distance market would be increased further.

To maintain the price difference between NTT and NCCs and, at the same time, to allow rate rebalancing by increasing effective local rates. The financial condition of NCCs would be worsened; the least competitive NCC, TeleWay Japan, might not be able to stay in business without assistance from outside. On the other hand, the share of NCCs in the long-distance market would continue to increase, and NTT would continue to suffer from losing its share.

Eliminate the price difference between NTT and NCC without introducing rate rebalancing. The rapid change in the share of the long-distance market between NTT and NCCs would be stopped.

It is yet to be seen which one of the three alternatives MPT will take or whether MPT will take some other alternative, such as elimination of the price difference accompanied by rate rebalancing.

ORGANIZATIONAL REFORM OF NTT

After the 1985 privatization, NTT carried out a number of major organizational reforms. They may be classified into two categories: external reforms (that is, creation of new organization such as subsidiaries; spinning out) and internal reforms. The objective of external reforms was to utilize NTT's resources, especially its labor force, more efficiently in various areas in which new business is emerging. The objective of internal reforms was to promote the internal business efficiency of NTT. Both external and internal reforms of NTT contributed to the recent increase in NTT's total factor productivity, a numerical measure of the performance of economic organizations such as NTT.[7]

External Reforms

The 1985 privatization of NTT gave it the possibility of creating subsidiaries. Soon after the privatization, NTT created a large number of subsidiaries in a wide range of business areas. MPT was generous, at least for a couple of years after the privatization, not to impose severe restrictions on NTT creating subsidiaries. The 1988 financial report of NTT states that the number of companies recognized formally as subsidiaries is 89 and the number of companies of which a part of the share is owned by NTT and which are under the influence of NTT is 70. The 1990 report gives the two numbers as 92 and 64, respectively.

The proportion of the share of a subsidiary owned by NTT ranges from 100 percent to 10 percent. The business of each subsidiary is related to some aspect of NTT's activities. For example, NTT Urban Development, a subsidiary, constructs office buildings on pieces of land owned by NTT in Tokyo. NTT Softwares, another subsidiary, develops software for computers used by NTT. NTT PC-Communications, Japan Information and Communication (NI+C), and NTT International, each being a subsidiary, utilize technology developed by NTT to promote computer communications. Other examples of subsidiary activities are operation of NTT's videotex (CAPTAIN), production and distribution of NTT's prepaid telephone cards, and sale of information obtained from NTT's yellow books. It is reported that by March 1989, approximately one-third of NTT's subsidiaries started producing surpluses but 20 percent of them were still in deficits.

The objectives of NTT's creation of these subsidiaries are to explore the possibility of developing new business by using NTT's resources outside MPT's regulation and to reduce the labor cost of NTT by transferring a part of NTT's work force to subsidiaries. During the one-year period preceding March 1989, approximately 3,500 employees were transferred "temporarily" from NTT to subsidiaries, 1 percent of the total number of employees of NTT in 1988. Creation of subsidiaries and transfer of employees was done without eliminating services provided by NTT, thus contributing to the increase in the total factor productivity of NTT.

In July 1988, the Department of Data Communications of NTT, its VAN (enhanced service) department, became independent and started as NTT Data Communications, Inc. The size of the new NTT Data Communications is approximately 10 percent of that of its parent. The objective for NTT to divest NTT Data Communications, Inc., was to let it move from the type I business to the type II business and operate outside the regulation by MPT. NTT Data Communications was by far the largest and the most efficient VAN firm in Japan; its services cover banking transactions, commodity distribution, credit authorization, and others. Many type II carriers opposed the separation of NTT Data Communications because the size and the efficiency of it, if not regulated, would be a blow to the business of type II carriers. MPT, on the other hand, was concerned with the direct interaction technological and business, between NTT as a type I carrier and NTT Data Communications as an entity doing type II business if they were allowed to remain as a single organization, and they approved the proposal of NTT for separation.

Internal Reforms

NTT started to reform itself to improve its efficiency soon after the possibility of privatizing and dividing NTT was suggested in 1982 by the Second Ad Hoc Committee on Administrative Reform. The committee stated that the need for privatizing and dividing NTT came from technological progress in the telecommunications industry and the presence of bureaucratic inefficiency within NTT, then a monopoly and a public corporation lacking incentives to improve itself.

Two major internal reforms of NTT are worth mentioning here. In 1985, NTT changed its internal organization from the departmental system to the profit-center system. Before this reform, NTT was organized according to functions of workers, such as telephone services, maintenance, procurement, construction, accounting and finance, research and development, and others. In 1985, NTT was reorganized into a profit-center system. Regional telephone centers were established; also created were the center for supporting telephone business, center for supporting corporate communication, center for advanced communication services, and center for networking.

Later in 1989, organization of regional telephone centers was simplified; the old three-level system was replaced by a two-level system, and terminal offices accommodating subscriber lines were turned into fully automated centers to which no worker is allocated. It is expected that this change will cut managerial and overseeing costs significantly.

Shortly before the privatization, NTT started to reduce the size of its labor force. For five years, beginning in 1984, the total number of employees was decreased by 8,174 per year, which was equal to 2.75 percent of its labor force in 1986. This means that NTT squeezed out 13.7 percent of its labor force in the five years. It was done by transferring workers temporarily to subsidiaries and by setting the number of new employees far below the number of retirees; no firing was done, as this is strongly against the Japanese employment practice. The speed of this change was impressive. MPT, however, recently stated that the saving of labor cost was not as high as was expected.

NTT also started, a couple of years before the privatization, what is called the ASK campaign ("ASK" stands for safety, quick response, and efficiency in Japanese). It is NTT's term for total quality control (TQC).

In Japan, TQC has long been utilized in many corporations as the most important strategy to increase the organizational efficiency. It is TQC that improved the quality of manufactured goods in Japan since the 1960s. TQC in Japan is carried out in the following fashion. Employees in a corporation are organized into groups, each composed of 5 to 20 employees working closely. The objective of a group is to find ways to improve the efficiency of the tasks done by the group. Efficiency is sought from many aspects: workmanship, coordination, technology, maintenance, scheduling, and others. Members of a group are encouraged to submit a proposal to improve the efficiency of the tasks of the group. Whether a proposal is adopted or not is determined jointly by the group members. There is a reward to the group that improves efficiency and also to the member of the group who makes a successful proposal. Furthermore, meetings and seminars are held within the organization to diffuse the information and the experience of successful proposals.

In NTT, ASK (TQC) started in 1983 under the leadership of Shinto, chairperson of NTT then, and spread to almost every section of NTT by April 1985. In 1987, the number of ASK groups in NTT was 97,000; approximately 90 percent of the workers of NTT participated in some ASK activities. It was the largest TQC campaign in Japan.

The ASK activities of NTT became the basis of improving NTT's efficiency from the bottom line of the organization. According to an expert, the cost saving due to its ASK campaign amounts to ¥1 billion annually, 0.2 percent of the annual gross profits of NTT in 1990.

CONCLUSION

To summarize, the 1985 reform of the Japanese telecommunications industry as, by and large, successful. First and foremost, because of the introduction of competition, the providers reduced the price of telephone services substantially and started offering a number of new services. There is no doubt that the reform was beneficial to users. Further, NTT benefitted by the reform, too. Because of the privatization, many of the regulations imposed on NTT prior to the privatization, then a public corporation, were removed; NTT now has more freedom to extend its activities over a wide range of business opportunities. In addition, because of the introduction of competition, NTT, which had been a sleeping giant before the reform, was reshaped into an active corporation, capable of growing in the competitive business environment today. Without the 1985 reform, NTT might have become an inefficient, bureaucratic entity like the old Japan National Railroad (JNR), which was bankrupted because of its failure to adjust to the new environment brought about by the development of automobile transportation and aviation.

The Japanese telecommunications industry today is not without problems, however. MPT, in 1985, apparently started to introduce an industrial structure into the Japanese telecommunications industry so that each competitive market would be operated by a small number of providers. The industrial structure is still in the state of transition. In the long-distance market, for example, the share of NTT is rapidly decreasing and the share of NCC rapidly increasing because of the price difference between NTT and NCCs imposed by MPT. MPT has not expressed a view as to whether the rapid change in the share should be stopped or, at least, slowed down. Assuming that this is affirmative, we do not know how MPT would do it. Should the price difference be eliminated, or should other actions be taken to stop or slow down the change in the share? A rapid change in the share between NTT and NCCs also can be seen in many of the telecommunications markets other than the local and the long-distance markets.

Such rapid change in the share has created uncertainty in the business environment of NTT. NTT, because of this, has started considering defensive actions that would damage NCCs if actually taken. MPT, therefore, is in the position to clarify what will be the "target share" of NTT and NCCs in each of the major telecommunications markets, implying that, on reaching the target, MPT will remove price differences. Whether MPT will actually do it is not known, but it is expected that something will be done by MPT to that effect and that the problem will be solved in some way in the future, perhaps with some confusion.

There is a long-run problem in the Japanese telecommunications industry that is more serious than the short-run problem stated above. It is the question "What is the desirable structure of the telecommunications industry in Japan?" It is widely agreed that there are two conflicting requirements in the telecommunications industry, unlike in industries such as manufacturing, agriculture, and many of the service industries. Let me call them the competition requirement and the public requirement. The competition requirement calls for introduction of competition into the industry so that the providers and the users can enjoy the benefits of competitive markets; it is now a classical thesis that requires no further explanation. The public requirement means that there are certain objectives that cannot be satisfied through competition: efficiency from large-scale operations (natural monopoly), universal services, standardization, and basic research for technical progress. These topics also have been discussed repeatedly and require no further explanation.

One of the most important issues in the telecommunications industry in Japan is where and how the two conflicting requirements should be reconciled. In 1984, the United States chose a divestiture of the (old) American Telephone & Telegraph (AT&T) to introduce competitive long-distance markets and monopolized regional markets. Japan did not break up NTT and introduced competition in the long-distance market. Thus, the Japanese telecommunications industry today is composed of competitive long-distance and other markets with regulated entry and a monopolized nationwide local market. NTT is allowed to operate as a monopolist in the local market and as a heavily regulated provider in the other markets. In 1989, MPT attempted but did not accomplish, to change this structure by dividing NTT. The question is to determine the future structure of the telecommunications industry in Japan.

One possibility is that the present structure will remain for some extended time. This seems to be the most likely scenario in view of the Japanese political background. NTT will be allowed to stay as a single entity but will be regulated heavily by MPT. Some of NTT's operations will be divested, one after another. NTT Data Communications was divested in 1988, and mobile telephone operations will become separate in 1992. Divestiture of the maintenance department and the department of satellite operations may follow in the future. The share of NTT in most of the competitive markets will be one-half or so, whereas in the local market NTT will remain the monopolist. NTT will suffer from the conflict of the private and the public requirements; it will be asked to be competitive and efficient on one hand and to behave as a public servant on the other. Consequently, a great deal of uncertainty will remain in decisions to be made by NTT and MPT on service provisions, pricing, and interfaces between providers.

The second possibility is the line of the 1989 MPT plan. This is not a likely scenario, because the Upper House of the Japanese Diet would oppose any attempt to divide NTT. In this scenario, NTT would be divided into at least two entities: NTT Long Distance and NTT Local, for example. The latter might further be divided into a small number of NTT Locals. This structure would be similar to the structure of the telecommunications industry in the United States today, except that in the Japanese long-distance market, entry is not free as in the United States, and it is regulated by MPT. Thus, if such division of NTT is made, the outcome will be similar to the one we see in the United States today. The long-distance market will grow rapidly, but the local markets will show differentiated development.

Local providers in metropolitan areas such as Tokyo or Osaka will grow rapidly. In addition to providing basic services, metropolitan local providers will have strong incentives to extend their operations both vertically and horizontally. Extension to equipment manufacturing, enhanced services, and international operations will be sought. Because of the division of the entire NTT, however, the overall level of research and development would be lower than otherwise. It would be difficult to achieve the objective of universal services; further, the advantage of operating under a small number of standards might be lost. Whether the potentiality of the telecommunications industry as a whole would be increased or decreased by dividing NTT is not known. Dividing NTT would decrease the average size of a provider, loosing the efficiency from large-scale operation. At the same time, however, dividing NTT would decrease the disadvantage of having bureaucracy of an excessive size, contributing to increasing the efficiency of operations of a provider. The classical controversy over the advantages and the disadvantages of centralization (or decentralization) would be appropriate to this issue.

The third possibility is to restructure the Japanese telecommunications industry according to rule(s) different than monopoly (the first possibility) or regional division (the second possibility). For example, functional or vertical division of NTT may be conceivable, if not likely. NTT may be divided into facilities (and maintenance) department, networking department, basic-service department, enhanced-service department, and so on. This scenario is very unlikely, but such a reform would partly solve the conflict between the private and the public requirements.

One issue that needs to be considered for future development of the Japanese telecommunications industry is the status of KDD. As the economic activities in Japan become more and more international, the size of the international telecommunications market will continue to grow. NTT, observing this trend, is seeking to extend its activities into the international business arena. In particular, NTT is seeking to integrate with KDD, although this is not being talked openly. KDD strongly rejects the idea and wants to stay independent. Probably MPT would not allow such integration, because it would contradict MPT's policy to decrease, not increase, the size of NTT. The managerial and technological resources held by NTT, however, are so large that it will be difficult to stop NTT from extending its activities overseas. Pressures will be mounted gradually if NTT is strictly confined to domestic activities. Something needs to be done in this regard.

Finally, one element affecting the future of the Japanese telecommunications industry is technological development. In particular, construction of broadband ISDN (B-ISDN), which will be the main issue in the industry during the second half of the 1990s, is crucial to the future structure of the industry, because B-ISDN, unlike conventional N-ISDN, will have to be constructed from scratch. This means that MPT has the freedom to design the telecommunications network in Japan in the twenty-first century. MPT has not yet expressed a clear view on this aspect, although it has disclosed reports on the technological and the commercial possibilities of B-ISDN.

NOTES

1. The Telecommunications Business Law of Japan allows type I carriers to engage in type II business (that is, to provide enhanced services). As of 1991, both NTT and KDD started, or are planning, to offer various enhanced services (value-added [VAN] services). NCCs, on the other hand, have not attempted to offer enhanced services except those that are directly related to the basic services they currently provide (such as credit-card calls). There is a possibility that "unfair competition" may emerge between NTT or KDD and type II carriers in providing enhanced services, because NTT and KDD own physical transmission circuits but type II carriers do not. In July 1970, on bilateral trade negotiations between the United States and Japan, the United States Trade Representative (USTR) and the Ministry of Post and Telecommunications (MPT) started a committee in which a framework similar to the open network architecture (ONA) in the United States would be considered to facilitate fair competition between NTT and KDD and domestic type II carriers in providing enhanced services.

2. Japan's Telecommunications Committee, "Interim Report on Japan's Telecommunications Industry in the Future," submitted to MPT, October 2, 1989.

3. Japan's Telecommunications Committee, "Report to the Ministry of Posts and Telecommunications on Regulations and Policies for Realizing Fair and Effective Competition and Promoting Technological Innovation — Actions to be Taken According to Article 2 of the Nippon Telegraph and Telephone Corporations Law," submitted to MPT, March 2, 1990.

4. As a consequence of the May 1989 U.S.-Japan agreement on allocation of radiowave frequencies, three, rather than two, mobile telephone carriers operate in the Tokyo metropolitan area. This was an outcome of the demand by the Motorola Corporation of the United States, determined to market its handy telephone terminals (Microtac) in Japan.

5. Japan's Telecommunications Committee, "Interim report."

6. Japan's Telecommunications Committee, "Report to the Ministry."

7. Hajime Oniki, Tae H. Oum, and Rodney Stevenson. 1990. The Productivity Effects of the Liberalization of Japanese Telecommunication Policy. Paper presented to the International Telecommunications Conference, Venice, March 1990. Mimeo. It reports a significant increase in NTT's total factor productivity which is attributed to the price reductions, the reduction of its labor force, and organizational reforms introduced by NTT.

REFERENCES

Business Week. 1990. "The Pacific Century." A *Business Week* symposium report, December 17, pp. 125–28.

Chen, H. 1990. "Plan for Developing Posts and Telecommunications." *China Market*, June, No. 6, p. 35.

China Communications Association. 1984. *China 2000: Status and Outlook of Telecommunications in China and the World.* Beijing: China 2000 Research project.

Greer, C. (Ed.). 1989. *China: Annual Facts & Figures,* Vol. 12, pp. 233–35. Beijing: Academic International Press.

Hao, W. 1990. Chief Engineer, Directorate General of Telecommunications, MPT. Personal interviews, August 1990.

Hauser, S. & Langhlin, M. 1991. "Market Entry Strategies in the Western Pacific Rim." Paper presented at Pacific Telecommunications Conference, Hawaii, January.

Ji, Y., et al. 1990. "Difficulty of Operations in Loss." *People's Posts and Telecom,* July 31.

Jing, X. 1986. "MPT Sets Ten Enterprise Groups." *People's Daily* (overseas edition), June 17.

Lerner, N. 1987. "Managing China's Complex Telecom Industry." *Telephony,* August 24, pp. 32–39.

Li, Y. 1991. "China to Double Number of Telephones to 23.8 Million in Five Years." *China News Digest,* January 25.

Liang, J. 1990. "Present State and Outlook of Optical Fiber Communications in China." *World Telecommunications,* August, Vol. 3, No. 3, pp. 8–10.

Liang, X. & Zhu, Y. 1988. "The Development of Telecommunications in China." Paper presented at Conference on Pacific Basin Telecommunications, Tokyo, September.

Ouko, R. 1987. "Criteria of Government Decisions in Developing Countries Concerning Telecommunications Investment." *Telecommunications Journal,* September, Vol. 54, No. 9, pp. 619–21.

Pierce, W. & Jequier, N. 1983. *Telecommunications for Development.* An ITU-OECD report. Geneva: International Telecommunication Union.

Saunders, R. 1983. "Telecommunications in the Developing World." *Telecommunications Policy,* December, Vol. 7, No. 4, pp. 277–84.

Song, Z. 1990. "China's Communications Towards Modernization." *World Telecommunications,* November, Vol. 3, No. 4, pp. 3–6.

Sun, L. 1990. "China Follows the Long Road to Telecom Growth." *Telephony,* May 28, Vol. 218, No. 22, pp. 22–30.

Sun, L. 1990. "Telecom in China: One Step Forward, Two Steps Back?" *TelecomAsia,* Winter issue, Vol. 1, No. 4, pp. 11–16.

Sun, L. 1991. "A Status Report on China's Telecommunications After 1989." Paper presented at Pacific Telecommunications Conference, Hawaii, January.

Townsend, D. 1991. "Telecommunications Infrastructure and Economic Development: Principles for Formulating Investment Policies." Paper presented at Pacific Telecommunications Conference, Hawaii, January.

Wellenius, B. 1984. "Telecommunications in Developing Countries." *Finance and Development,* September, Vol. 21, No. 3, pp. 33–36.

Wellenius, B., et al. 1989. *Restructuring and Managing the Telecommunications Sector.* Washington, D.C.: World Bank.

World Bank. 1990. *World Development Report 1990.* Oxford: Oxford University Press .

World Telecommunications. 1989. "China Telecommunications Services and Investment in 1988," May, Vol. 2, No. 2, p. 53.

World Telecommunications. 1989. "Inspection for SPC PBXs," August, Vol. 2, No. 3, pp. 23–26.

World Telecommunications. 1989. "The Use and Development of Local Telephone Exchanges," November, Vol. 2, No. 4, pp. 10–12.

World Telecommunications. 1990. "China Telecommunications Services and Investment in 1989," May, Vol. 3, No. 2, p. 55.

Zhang, P. 1986. "Outlook of Posts and Telecom in 1986." *People's Daily* (overseas edition), February 5.

Zhang, T. 1990. "Basic Telecommunications Statistics in China." *World Telecommunications*, February, Vol. 3, No. 1, p. 53.

Zhou, C. 1990. "Strengthening Management and Development of Communications Industry." *World Telecommunications*, February, Vol. 3, No. 1, pp. 3–4.

6

Telecommunication Reform in Canada

William Melody and Peter S. Anderson

Telecommunication has played a prominent role in Canadian history, and Canada has played a prominent role in telecommunication history. Alexander Graham Bell may have spoken the first words over the telephone, "Mr. Watson, come here, I want you," in Boston on March 10, 1876, but five months later he undertook the first definitive tests of the telephone in Brantford, Ontario, including the first one-way long-distance call over a 13-kilometer line he had built between Brantford and Paris, Ontario. The construction of local telephone systems began immediately thereafter in cities and towns across North America. Eventually, Bell moved to Nova Scotia, where he continued his inventive activity and retired.

More recently, Canada was the first country to have a domestic satellite system (1973). Included in its pioneering activities are the development and application of services to remote areas, certain aspects of network digitalization, and innovations in satellite and space communication.

In the 1870s, the Canadian government extended telegraph lines across the Plains as part of its program for staking a claim against possible U.S. expansion northward. As Canada is a nation in which 80 percent of the population lives within 100 miles of a 4,000-mile southern U.S. border, most Canadians are far closer to the United States than they are to most other Canadians. Natural geographic and economic links for most Canadians have been stronger in the north-south direction than east-west. Moreover, Canada's major population growth has been driven by waves of immigration rather than indigenous growth of a population with strong historical ties to the land.[1]

In a vast land where a small population is spread across a considerable amount of remote and rugged terrain, with harsh weather conditions for a significant portion of the year, communication has always ranked high on Canada's political, economic, and social policy agenda. The telephone penetration rate is nearly universal and one of the highest in the world. On a per capita basis, Canadians are among the most active communicators over the telecommunication network. The Canadian broadcasting system is nearly universal. Canada was a world leader in the widespread introduction of cable television. By necessity, communication has been central to the development and maintenance of Canada as a nation. Telecommunication has been an important part of this development and remains so today.

This chapter provides an overview of the historical development of telecommunication in Canada and the major policy issues currently under investigation and their potential implications. Like all countries, Canada is currently in a process of telecommunication reform, but because the performance of its telecommunication sector has been very good by world standards, Canada's approach to reform has tended to reflect more-modest changes to update a satisfactory system than watershed reforms of inefficient and unresponsive systems. Nevertheless, its proximity to the United States means that Canada's telecommunication future is likely to be influenced as much by U.S. as Canadian policy.

Telephone development began in 1879 with the chartered Bell Telephone Company of Canada.

Between 1950 and 1975, Bell Canada acquired most of the larger and growing independent companies in its region during a period of industry consolidation. In 1976, the federal regulatory responsibility for telecommuncations was transferred to the Canadian Radio Television and Telecommunications Commission (CRTC).

Throughout the 1980s, there was a limited degree of competition for long-distance services from an increasing range of overlapping services with the Canadian National and Canadian Pacific (CNCP) telegraph company; some new, enhanced, or value-added service suppliers coming on to the network; and some Canadian businesses bypassing the Canadian network by integrating their demands into U.S. systems. This was facilitated by the fact that the majority of Canadian industry reflects branch plants of U.S. firms or Canadian companies with strong links to U.S. suppliers or customers.

With rapid cable television development in the 1970s, the possibility of competition between telephone companies and cable companies for the supply of some telecommunication services became an issue, but government policy did not encourage it. After much debate about the applicability of a forced separation of content and carriage, a policy of separation was not adopted.

Under the Canadian regulatory system, the federal government had jurisdiction over only telecommunication carriers in regions that it had licensed historically, that is, primarily Ontario, Quebec, and British Columbia. Although this represented only three of ten provinces, it represented approximately 75 percent of the Canadian population. However, provincial regulations in other provinces had an impact on the development of national network systems and services.

After generations of controversy, in 1991, the Supreme Court of Canada awarded the federal government complete jurisdiction over all interprovincial telecommunication services. In 1991, the CRTC conducted major hearings on the application of Unitel for a license to establish a sound national network of public long-distance telecommunication services. If it is accepted, it will introduce a degree of direct competition in Canada paralleling second carrier policies adopted in the United Kingdom and Australia.

INDUSTRY STRUCTURE

The Established System

The Trans-Canada telephone system (TCTS) was established as an unincorporated association of the larger privately and publicly owned telephone companies across Canada to coordinate the planning and development of the national network. It also markets national services and products and distributes network revenues among the participating carriers. In recent years, TCTS changed its name to Telecom Canada.

Telesat Canada, the domestic satellite carrier — jointly owned by the government of Canada and the Canadian telecommunication carriers — joined Telecom Canada in 1977 to participate in the planning and investment decisions affecting network development and to facilitate integration of satellite facilities into the terrestrial systems operated by the other members. In the early years of Telesat's operations, the other members of Telecom Canada guaranteed the financial liability of Telesat through transfer payments. In recent years, Telesat has sought to get out of these arrangements in order to deliver services directly to subscribers across Canada, essentially in competition with the other members.

Telecom Canada members provide a range of facilities for the transmission and switching of local and long-distance telecommunications traffic, including two coast-to-coast microwave relay routes. Telecommunication traffic is also carried on coaxial and fiberoptic cable systems and by Telesat satellites. Table 6.1 lists the members of Telecom Canada. Recent financial statistics of Canadian carriers are provided in Table 6.2. In addition to the Telecom Canada members,

TABLE 6.1

Telecom Canada Membership

Carrier	Ownership	Principal Territory
Bell Canada	Private	Ontario, Quebec, and eastern portion of Northwest Territories
British Columbia Telephone Company	Private (U.S. GTE Co.)	British Columbia
Alberta Government Telephones	Recently privatized	Alberta
Saskatchewan Telecommunications	Public	Saskatchewan
Manitoba Telephone System	Public	Manitoba
Maritime Telegraph and Telephone	Private	Nova Scotia
New Brunswick Telephone	Private	New Brunswick
Newfoundland Telephone	Private	Newfoundland
Island Telephone	Private	Prince Edward Island
Telesat Canada	Public/Private	Canada

Source: Canadian Radio-television and Telecommunications Commission, *Annual Report, 1989–1990*.

TABLE 6.2

Selected Financial Statistics of Canadian Carriers, 1989
(in millions of dollars)

Carrier	Operating Revenue	Net Income	Gross Plant	Total Assets
Bell Canada	7,273	875	20,915	15,699
British Columbia Telephone Company	1,689	181	4,575	3,453
Maritime Telegraph and Telephone	454	49	1,422	1,139
New Brunswick Telephone	305	33	886	615
Newfoundland Telephone	250	32	771	599
Teleglobe	223	34	488	703
Telesat Canada	147	23	1,075	781
Island Telephone	48	5	155	117
Total Industry	13,692	1,465	40,327	30,364

Source: Canadian Radio-television and Telecommunications Commission, *Annual Report, 1989–1990*.

there are still approximately 100 independent telephone companies operating across Canada.

Teleglobe Canada carries overseas telecommunication traffic through its international gateway switches at Vancouver, Toronto, and Montreal. It uses undersea cables jointly owned with international carriers of other countries and satellite facilities leased from Intelsat. Traffic to and from Canadian points is generally carried by Canadian domestic carriers to gateway switches and thereafter on Teleglobe facilities. Services to the United States and Mexico are provided by Telecom Canada members, not Teleglobe.

In 1987, the ownership of Teleglobe was transferred from the government of Canada to Memotec Data Inc. Coincident with the sale was a government policy to extend Teleglobe's effective monopoly of overseas traffic for a minimum five-year period. These arrangements are now under review. The CRTC has also initiated a proceeding to determine the extent to which Teleglobe should be regulated.

Unitel

Unitel Communications Inc., because of its unique history of supplying telegraph-related services, is a national facilities-based carrier. It provides a wide range of competitive private voice, data and messaging services (excluding public switched, local, and long-distance telephone services) across Canada. Unitel's predecessor CNCP was granted the right to interconnect to the local distribution facilities of Bell Canada in 1979 and British Columbia Telephone in 1981 for the provision of competitive private voice and data services. Unitel operates its own national microwave relay system and switching centers and has fiberoptic systems installed on major high-density routes as part of an eventual Canada-wide fiberoptic network. Unitel's major owner, Rogers Communications, is the dominant cable television owner in Canada. In addition, Rogers has holdings in Canadian radio and television, pay television (TV) and cable specialty programming services, and video rental stores, as well as interests in radio systems.

Mobile Communications

Radio common carriers perform in a small but rapidly growing sector of the industry. In 1963, the first licenses were granted to provide radio paging, mobile radio, message forwarding, alarm monitoring, and data transmission services. In 1985, the government licensed cellular radio telephone service on a duopoly basis, to be provided by the established telephone companies and an independent private operator, Rogers Cantel Inc. Cantel provides service to more than 24 Canadian centers.

In recent years, a growing number of enhanced service providers, private network operators, and resellers have come into niche sectors of the Canadian market. However, the extent of liberalized access to the Canadian network has not been as great as that of the United States, because the public network services have been fully protected from the impact of competition in the fringe services.

CANADIAN TELECOMMUNICATION POLICY AND REGULATION

The Canadian regulatory environment[2] has evolved from a structure in which telecommunications common carriers were regulated by municipal councils, provincial cabinets, provincial government regulatory bodies, or a federal regulatory agency. Currently, the federal government and its regulatory agency, the CRTC, hold the dominant policy- and decision-making positions, but federal jurisdiction over all telecommunication sectors is not absolute. Until very recently, carriers in Newfoundland, Nova Scotia, Prince Edward Island, New Brunswick, Manitoba, and Alberta were regulated by their respective provincial public utility boards or commissions.

However, by a Supreme Court of Canada constitutional ruling in 1989, federal jurisdiction was extended over these carriers, with the exception of provincially owned carriers. By subsequent changes in ownership and federal/provincial agreements, all but SaskTel are now federally regulated.

The remaining independent telephone companies are regulated at the provincial level in British Columbia, Ontario, and Quebec. Edmonton Telephone and Prince Rupert City Telephones are both regulated by their respective municipal councils.

By legislative changes and federal/provincial agreements, Canada is moving toward establishing an integrated national communication policy framework, with greater provincial involvement in its administration. Also, changes recently approved in federal broadcasting legislation and proposed amendments to federal telecommunications legislation are aimed at expanding the powers of the federal cabinet to give direction and set broad policy goals.

Under present arrangements, the CRTC regulates all the Telecom Canada affiliates except SaskTel, Northwestel, Teleglobe Canada, and Terra Nova Telecommunications. Federally regulated carriers account for more than 78 percent of the telecommunication service revenues, 83 percent of net income, and approximately 83 percent of the total assets of the Canadian telecommunications industry.

Federal Telecommunications Regulation

The CRTC's authority to regulate telecommunications is derived from various statutes. The principal one is the Railway Act. Under the Act, the Commission is required to approve tolls prior to their implementation, based on finding that the rates are just and reasonable, are nondiscriminatory, and do not confer an undue preference. The Act also requires Commission approval of any interconnection and operating agreements.

In weighing the public interest, the CRTC attempts to canvass a wide range of potential benefits and disadvantages that might arise in granting an application, including those presented by the applicant, respondents, or any intervenors. In addition to considering the justness and reasonableness of rates, the Commission also considers such factors as universality of service; consumer choice and responsiveness to consumer need; quality of service; innovation in the Canadian telecommunications industry; efficiency of the telecommunication systems; optimum allocation of resources, taking into account geographic differences; the structure of rates, including route-averaged pricing, rate group structures, and rural service rates; and industry structure.

Taking account of these factors, the Commission has applied the Railway Act in different ways to different companies and services. The telephone companies are subject to rate-of-return regulation and rate regulation of individual services. Unitel and Telesat are also subject to rate regulation. The Commission's practice with respect to cellular companies is to grant expedited interim approval based on a prima facie finding that the proposed rates are just, reasonable, and nondiscriminatory. The Commission has declined to regulate in any way the services of resellers, having determined that they are not companies within the meaning of the Railway Act. This inaction, however, recently has been challenged in the courts by the Telecommunications Workers Union, with a decision yet to be handed down.

Liberalization of Canadian Telecommunications

A driving force behind many of the jurisdictional changes discussed above is the federal government's desire to consolidate its authority over telecommunication in order to establish a national telecommunication competition policy. By a series of CRTC regulatory and federal Department of Communication spectrum licensing decisions, competition has been permitted in a number of sectors, including terminal equipment, paging and mobile services, cellular radio, data services, enhanced services, local system interconnection, interconnected private line voice, private line services, and resale. These actions are summarized below:

1977 The CRTC holds that customers can own automatic mobile radio equipment in order to access the Public Switched Telephone Network.

1979 The CRTC authorizes Unitel (then CNCP) to connect its private line intercity voice and data services and facilities with the local exchange facilities of Bell Canada's public switched telephone network in Quebec and Ontario in order to provide competitive interconnected private services in Bell's operating territory. The Commission noted that as a matter of regulatory policy, all interexchange services that are directly competitive and that co-use local exchange facilities should make a comparable level of contribution toward local exchange facilities costs, at levels to be determined by the Commission.

1980 Terminal attachment approved on a final basis.

1981 Unitel is authorized to connect its private line intercity voice and data services and facilities with British Columbia Telephone Company's public switched network on the same terms and conditions as Bell Canada's network.

1983 The Department of Communications announces that a radio license will be issued to Cantel to provide service in competition with telephone companies across Canada.

1984 The CRTC permits cellular operators and conventional public and private mobile radio systems to carry interconnected long-distance traffic on their networks, provided that it is either originated or terminated on a mobile terminal. In many territories, cellular carriers can carry cellular Public Message Toll Service (PMTS) traffic on their own systems as long as the traffic either originates or terminates on a portable terminal. No contribution charge is applied.

The CRTC establishes the basic rules for distinguishing between enhanced and basic services and removes restrictions on resale and sharing to provide competitive enhanced services.

1985 The resale of private line services and sharing is authorized. The CRTC noted that entry by resellers into a market in which rate differentials substantially exceed cost differentials could result in uneconomic entry and place the carriers at a competitive disadvantage.

To minimize these problems, the Commission considered the use of contribution charges. It was decided at that time, however, that because of the variety of resale and sharing activities that could be developed, any contribution mechanism would be impractical. With regard to resale and sharing to provide PMTS and Wide Area Telephone Service (WATS), the CRTC noted that in this market, the rate differential greatly exceeded the cost differential, and because it had found that contribution mechanisms were inappropriate, the Commission disallowed such resale and sharing.

However, with respect to resale and sharing to provide non-PMTS/WATS interexchange services, the Commission noted that the discrepancy between rates and cost was not significant and that entry, therefore, could be allowed without the requirement of contribution payments.

1987 The CRTC approved tariff revisions that allow the resale and sharing of carrier services to provide local services (except public pay telephone services), data services, and voice services other than PMTS and WATS. This decision clarified the restrictions under which resellers could operate. Specifically, it required resellers to provide dedicated facilities to their customers. This made it possible for resellers to offer discounts on U.S. and overseas traffic, because only the Canadian portion had to be dedicated to individual customers. Consequently, resellers focussed their marketing on the U.S. and overseas traffic of large business users.

1990 The CRTC allowed the resale and sharing of private line services for joint use and removed the restriction that unsold facilities had to be dedicated to individual customers. The Commission also authorized the provision of interconnected interexchange voice services.

With respect to contribution charges, the Commission concluded that the imposition of a contribution charge of an appropriate level would provide the necessary mechanism for permitting such resale and sharing while limiting opportunities for uneconomic entry.

With respect to competition in public long-distance telephone service in Canada, so far, telephone companies have been permitted to retain a monopoly in PMTS and WATS. However, these arrangements are now under challenge.

The Regulatory Process

Canada has developed procedures for information gathering, public input, and critical analysis to implement a public interest in the policy-making process. By way of illustration of this process with respect to a major policy issue, the CRTC's approach to examining the Unitel application to become a second telecommunications carrier in Canada is described below.

Section 69 of the National Telecommunications Powers and Procedures Act authorizes the Commission to "make general rules regulating its practice and procedures." Under this authority, the CRTC has established a set of Telecommunication Rules and Procedures.

Consistent with its Rules and Procedures, the CRTC in May 1990 set out procedures as to how it would proceed with considering the Unitel application. On August 3, 1991, it refined these procedures.

Participants were invited to comment on the advantages and disadvantages of alternative scenarios to reduce the price of long-distance service, including the applications by Unitel; further liberalizing the rules for resale and sharing; rate rebalancing; and the entry of multiple long-distance voice competitors. However, the CRTC ruled that during the proceeding, it would not consider specific rate rebalancing proposals. It also noted that without causing unacceptable delays in the public process, it would be very difficult, if not impossible, to include any additional applications for long-distance service. The Commission asked participants to examine and comment on the impact the different scenarios for lowering long-distance rates would have on the revenues of telephone companies named in the Unitel application; rates for interexchange services and route averaging; affordability and accessibility of local service; urban and rural subscribers in different regions; telephone companies' obligation to provide service; other telephone companies operating in Canada; telecommunications network planning and design; service quality, choice research and development (R&D), and innovation; international competitiveness; and bypassing of Canadian network facilities.

The CRTC also outlined a number of topics related to the regulatory regime that it wished to examine, including the general rules governing treatment of competitors and existing telephone companies and the type and level of contribution payments or other regulatory mechanisms, if any, that would be necessary to ensure the maintenance of universally affordable service.

The CRTC invited the participation of federal, provincial, and territorial departments and agencies concerned with telecommunication regulatory policy and of telecommunication common carriers that were not respondents in the proceeding, a well as other interested persons and organizations. The basic procedures for participating in the proceeding are described below.

Persons wishing to receive copies of Unitel's applications, replies, and evidence or the answer and evidence of any Unitel respondent were asked to file such a request with the parties concerned.

Between February 25, 1991, and April 15, 1991, the Commission scheduled a series of 12 informal regional hearings across Canada to give Canadians a chance to give the CRTC their views on the applications and associated issues. The submissions were heard by two CRTC Commissioners who were part of the Central Hearing panel. Unitel was directed to send senior representatives to the hearing to provide points of clarification. A transcript from these hearings, along with filed documentation, forms a part of the official record of the entire proceeding. Further, for those unable or not wishing to attend these hearings written comments to the CRTC would be accepted at any time until the end of the central hearing. Written comments were to be treated the same as what was said in person at the hearings.

The formal Central Hearing commenced on April 1, 1991, in Hull, Quebec (adjacent to Ottawa). Persons (intervenors) wishing to participate in this hearing were required to file notice of intent to participate with the CRTC, with a copy sent to applicants and respondents. Unitel, respondents, and intervenors were considered parties to the proceeding.

Unitel was directed to serve on intervenors copies of their applications, particulars, replies, and evidence. Similarly, respondents were directed to serve on intervenors copies of their answers and responses to interrogatories. Parties were then given an opportunity to address interrogatories to Unitel with respect to its application and its filed evidence, with Unitel to serve responses to all parties. Parties could then file requests for public disclosure of information for which confidentiality was claimed, responses to their interrogatories, and requests for further responses, with appropriate reasons. Replies to such requests were then to be made by the applicants and served on all parties.

On February 15, 1991, the Commission issued a decision with respect to requests for additional disclosure and, on April 8, 1991, held a prehearing conference to deal with remaining matters and to establish the framework for the April 15, 1991, Central Hearing. Its April 10, 1991, Prehearing Conference Decision set out procedures for accepting opening statements and examination-in-chief, the ordering of parties presenting evidence and cross-examination, sitting schedule, and presentation of final argument.

Hearings were completed by the end of June. There remained the possible filing of rebuttal evidence, the filing of written briefs by the parties, and oral agreement before the Commission issued a written decision. This entire process should have been completed by the end of 1991.

TELECOMMUNICATION LIBERALIZATION IN THE CONTEXT OF CANADIAN ECONOMIC POLICY

Although the Canadian government has not attempted to implement a formal national economic policy with respect to the telecommunication industry, there has been a consistent recognition that a Canadian-controlled telecommunication infrastructure has been vital to the future growth and development of the Canadian economy, even in recent times. For example, in 1979, the Consultative Committee on the Implications of Telecommunications for Canadian Sovereignty urged "the Government of Canada and the governments of the provinces to take immediate action to establish a rational structure for telecommunications in Canada as a defence against the further loss of sovereignty in all its economic, social, cultural and political aspects."[3] With respect to satellite communication, the Department of Communications stated in 1980 that "[o]bviously the availability of these [satellite] services will be an essential factor in the continued social and economic development of the country. Control of the facilities and data flow will be an important consideration in the maintenance of our cultural and economic sovereignty."[4]

The telecommunication infrastructure in Canada must grow in a way that is responsive and complementary to broader aspects of national economic policy. As international markets for Canadian products have become more competitive, it has become necessary for the Canadian economy to diversify and to specialize in the production of selected high-technology products. To achieve broad economic objectives that require continuing growth and profitability and increased employment opportunities, the telecommunication and information services that provide the network for the flow of both domestic and international information in Canada must facilitate communication on an east-west basis and between Canadian urban centers and distant manufacturing and production centers.

A priority for increasing the international competitiveness of Canadian-produced goods and services is to lower the costs and enhance the quality and diversity of communication services within and between firms located in Canada. Financial and other forms of information about domestic markets and production priorities within Canada are as important, if not more so, as access to information regarding international markets. If the Canadian economy is to be strengthened by its telecommunication network, it must be able to provide a full range of voice and data services to all sectors of the Canadian economy at rates that do not inflate production costs above those of international competitors. Therefore, it is important to recognize the impact of U.S. competition policy on the Canadian telecommunication market, on national economic objectives, and on Canadian control over its own economy.

THE IMPACT OF THE U.S. COMPETITION POLICY ON CANADIAN TELECOMMUNICATIONS

In the transition from monopoly to competition, Canadian policy is at least 15 years behind the United States. The United States is in the final stages of the transition. The surviving new competitors have grown for more than a decade. The established telephone companies gradually have reoriented their activities away from erecting artificial barriers to new competitors. They have begun to act more like competitors than recalcitrant monopolists. The struggle between the old monopoly policies and practices and the new competitive ones is over. All parties have accepted the fact that the new competition has established itself and that there is no turning back to the nostalgic days of government-protected monopoly.

The adjustment process has not been easy for many companies, especially the former telephone monopolies and the historic token competitor, Western Union. With the American Telephone & Telegraph (AT&T) divestiture, both AT&T and the Bell Operating Companies have cut back on their labor and management forces by 15 percent to 20 percent. AT&T has taken a substantial obsolescence asset write-off that although primarily related to terminal equipment, included write-downs of the value of some equipment used for network services. Aggressive marketing activity and diversified service offerings responding to the specialized needs of different industries and types of use have replaced the traditional passive and reactive industry approach to marketing and services. The telephone companies have found it necessary to replace telephone sales and marketing people with recruits from the computer industry. Price competition has been vigorous. Western Union, which historically occupied a similar position in the U.S. market to that of CNCP in the Canadian market, went to the verge of bankruptcy and was bought out by a competitor. The most successful new competitor, MCI, has expanded its markets to include overseas services and joint services to Canada and Mexico. AT&T has become a much more aggressive competitor in international markets and has entered Canada to the extent permitted by Canadian policy.

The U.S. long-distance companies ready to enter the Canadian long-distance market in a big way include AT&T, MCI, GTE-Sprint, and several enhanced service suppliers. In addition, at least three or four resellers could easily extend their coverage to include the larger Canadian cities. There is a large attraction because Canadian long-distance rates are significantly higher than U.S. rates.

The most efficient way for U.S. companies to extend their facility networks would be to connect Canada's larger cities to the nearest border point of their U.S. systems, for example, Vancouver-Seattle, Toronto-Buffalo, or perhaps from the top of the Canadian National

(CN) tower in Toronto across Lake Ontario. This could be done with a few microwave hops, miles of fiberoptic cable, or satellite dishes. Probably no more than one or two companies would build facilities, and if existing capacity is available at reasonable prices from a Canadian carrier, such as CNCP, a railroad, a hydro company, or Telesat, they could lease it.

In most cases, other Canadian cities would be served, if there was enough demand, by having them connect to the nearest major city. However, some would link up directly to a U.S. city, such as Detroit, which would connect with Windsor. Facility extensions could be made — most probably by radio microwave — in a few instances, such as Ottawa and London, where demand would be high. The rest of the nation could be served using leased lines and resold WATS from the established telephone companies to the extent that the various cities and towns could be profitably served.

With this system, there would be relatively little new facility construction in Canada because of the economies of adding Canadian demand to the existing U.S. facility networks. Montreal-to-Vancouver traffic could be served efficiently over the U.S. system. Moreover, the dominant traffic patterns involving Canada in North American telecommunication networks are north-south, primarily because of the branch plant character of most Canadian industry. Addition of the entire Canadian long-distance market could increase demand on the U.S. system by about 10 percent, less than a year's growth. It would be less than half the unused capacity on the system at any point in time. Obviously, the Canadian demand would not be uniformly distributed over the U.S. system, but when considering alternative routing possibilities and the enormous capacity of the newest cable (fiberoptics) and radio microwave systems, absorption of the Canadian demand between major cities would not be a difficult task. Indeed, recognition of this possibility may have been a significant factor in bringing acceptance in Canada of the need for a comprehensive national policy-making authority.

In contrast to the aggressive competitive marketing of telecommunication services in the United States, the Canadian market is still highly regulated. It has accommodated a limited degree of "gentlemen's" competition gradually over time in a manner that has not seriously threatened any established interests. To date, the CRTC's view of competition has restricted serious price competition. Resistance to competition remains strong and quite successful. Preservation of the shared monopoly option is still very much the dominant policy. Neither the industry nor the regulators are prepared for hard-nosed competition, U.S. style. In sum, the U.S. companies are now finely tuned competitors ready to extend their markets. The Canadian companies and policy makers are busy jockeying for position and dealing with problems in the Canadian market. Their range of vision has extended, for the most part, no farther than the border. The door to international competition has swung inward for foreign competitors. There are few signs of Canadian companies going the other way.

The benefits of U.S. competition for Canadians would be in the significantly lower rates and the more diversified service offerings available. Ironically, the major beneficiaries are likely to be smaller business users and residential users. Large business and government users and special service users (for example, data) now receive substantially reduced discriminatory rates through special tariffs. Competition tends to eliminate monopolistic price discrimination by bringing down the prices of the higher-priced services. Therefore, those already receiving discriminatory low rates are likely to benefit less than those who are not. In any event, long-distance rates in Canada and between Canada and the United States could be expected to decline significantly.

The response of the Canadian telephone companies, because they supply both long-distance and local services in common, will be to raise local telephone rates to compensate for competitive rate reductions in the long-distance market. The U.S. telephone companies have responded to long-distance competition by imposing special charges on telephone subscribers for access to the telephone network for any service at all — the customer access line charge.

Bell Canada already has submitted a study to the CRTC claiming that if all the local loop costs that are common to local and long-distance service are allocated to local service, the costs of local service would be $30–$40 per month. If competition is permitted in long-distance, Bell Canada and B.C. Tel would propose to allocate their costs in this manner. Nevertheless, the major impact of competition from U.S. carriers is likely to be a major restructuring of rates by the Canadian telephone companies, reducing long-distance and increasing local rates, to make entry by the U.S. carriers less profitable.

There is a significant probability that Telesat, Teleglobe, and (possibly) Unitel (if its application is approved) could all be put in a deleterious financial position by active competition from U.S carriers. The government may be faced with an important policy decision as to whether it wishes to subsidize any of these companies to maintain a stronger Canadian presence in the market.

Northern Telecom (Nortel; Canada's leading transnational corporation in the telecommunication equipment field) has been very successful in the U.S. and international markets and should be expected to do well in a competitive Canadian market. Nevertheless, it presently has major captive customers for several lines of telephone equipment used by Bell Canada and its Maritime subsidiaries. Competition should force some of these markets open, causing a potential loss in market share to Nortel. Perhaps even more significant for the long term may be an increased incentive for Nortel to move its headquarters to the United States if it loses its privileged position in the Canadian market and if it can obtain the full benefits from competing in Canada even though it is no longer a Canadian company.

Teleglobe's privileged position as the Canadian monopoly on overseas communication could also be threatened. AT&T, MCI, and the other U.S. carriers have their own overseas connections and services. There has been competition among U.S. companies for years, and it has become quite aggressive. Not only could the U.S. companies carry their own Canadian customers' overseas traffic, they also could likely improve on Teleglobe prices and revenue-sharing agreements with Bell Canada and other Canadian telephone companies that originate and terminate Canadian overseas traffic. Teleglobe's days as a high-profit monopoly clearly soon will be over, and unless it can adjust to competition instantly, it too could be placed in financial difficulty.

U.S. competition could require major changes in the current structure of Telecom Canada and could lead to its demise. The current Revenue Settlement Plan (RSP) by which the telephone companies share the revenues from providing cross-country traffic is based upon a presumption of cooperative monopoly sharing. Under competition, this system would have to be completely reorganized and a new method worked out in which local telephone companies charge long-distance carriers for obtaining access to customers via local company exchange facilities.

U.S. competition would promote certain patterns of communication and discourage others. Rates between Canada's largest cities would probably decline, as would rates to U.S. cities. North-south communication patterns would be encouraged by major rate reductions. Local and most intraprovincial communication patterns would be discouraged by increasing rates. Canadians may be able to obtain access to U.S. data banks, value-added services, and information services at lower cost than Canadian alternatives. This would help Canadians as consumers of these services but would discourage Canadian companies attempting to establish these services in Canada. The Canadian data processing industry estimates that Canada already has lost more than 200,000 computer software jobs because of transborder data flow to the United States.

The problem of Canadian control over Canadian content, an historic concern of Canadian governments, would be exacerbated for all types of content. Data and information banks would acquire a status similar to that of direct broadcast satellite (DBS) delivery of TV signals today. Canadians would have easier access to U.S. services through north-south connections than to equivalent Canadian services.

The impact on Canada's balance of payments could be substantial. Competition from U.S. telecommunication carriers will make it easier for Canadians to obtain access to U.S. goods and services, placing Canadian suppliers of similar goods and services at a competitive disadvantage. Under the Canada-U.S. Free Trade Agreement, telecommunication services would provide efficiency and service advantages for the benefit of all U.S. industry in entering Canadian markets and impose a competitive disadvantage on Canadian firms operating in and attempting to enter U.S. markets. It is difficult to see a Canadian carrier making a successful entry to the U.S. market or the competitive advantage of any Canadian industry being improved, except for the retailing of U.S. services. Canada's future in the information economy would tend to be even more dependent on the U.S. economy, making the unresolved questions relating to Canadian control over Canadian information, privacy, and so on even more serious than they are now.

CONCLUSION

The Canadian telecommunication market, supplied by a cartel of regional monopolies and protected by regulatory policies that discourage aggressive competition, may soon be facing a substantially altered market environment. The U.S. carriers, honed by a decade of aggressive competition and industry restructuring, are showing signs of expanding significantly into the Canadian marketplace. The Canada-U.S. Free Trade Agreement and developments at the General Agreement on Tariffs and Trade (GATT) liberalizing trade in the services sector are reducing barriers to entering the Canadian market. As consumers of long-distance telecommunication and information services, Canadians in the major cities could benefit significantly by low-cost access to U.S services. However, most local and intraprovincial rate are likely to increase significantly, and Canadian suppliers of telecommunication and computer equipment and services, as well as information and data bank services, could suffer. Some jobs could be lost, and the trade deficit in the information sector would be increased.

However, Canada can put off the day of reckoning with the microelectronics revolution and its implications for the information economy only so long, and the longer the delay, the more difficult will be the adjustment. Canada can make the adjustment in a manner best for Canadians by encouraging complete liberalization and real competition among domestic suppliers immediately. In a sense, it would be ironic for Canadian policies to place greater restrictions on Canadian entrepreneurs than on U.S. firms. Many imaginative Canadians have been restricted from pursuing their proposals in telecommunications because of the historic monopoly policies. If

Canadian entrepreneurs are turned loose to improve the efficiency of the telecommunication industry in Canada, Canadians can benefit in all respects.

Only after an adjustment period of several years are Canadian firms likely to be in a position to hold their own against competition from the U.S. carriers. The adjustment period has only just begun in Canada, and at the current rate of change, it is not likely to progress significantly until events require it. The government could start the adjustment process now by undertaking major policy initiatives along the lines suggested above.

NOTES

1. As in many countries developed by conquest, the native population has been marginalized and is still fighting for basic rights.

2. Peter S. Anderson, *Unitel Communications Inc. and B.C. Rail Telecommunications/Lightel Inc. — Applications to Provide Public Long Distance Voice Telephone Services and Related Resale and Sharing Issues: A Field Report.* Discussion Paper (Melbourne: CIRCIT, 1991).

3. Consultative Committee on the Implications of Telecommunications for Canadian Sovereignty, Telecommunications and Canada (Department of Communications, Ottawa: Supply and Services, March 1979).

4. Department of Communications, *The Canadian Space Programme: Five Year Plan (80/81–84/85)* (Ottawa: Supply and Services, 1980).

REFERENCES

Peter S. Anderson, *Unitel Communications Inc. and B.C. Rail Telecommunications/Lightel Inc. — Applications to Provide Public Long Distance Voice Telephone Services and Related Resale and Sharing Issues: A Field Report.* Discussion Paper (Melbourne: CIRCIT, 1991).

Robert E. Babe, *Telecommunications in Canada* (Toronto: University of Toronto Press, 1990).

W. G. Bolter, J. B. Durall, F. J. Kelsey, J. W. McConnaughey, *Telecommunications Policy for the 1980's — the Transition to Competition* (Englewood Cliffs, N.J.: Prentice Hall, 1984).

Gerald W. Brock, *The Telecommunications Industry, The Dynamics of Market Structure* (Cambridge: Harvard University Press, 1981).

Robert J. Buchan et al., *Telecommunications Regulation and the Constitution* (Montreal: IRPP, 1982).

Canada, Department of Communications, *Canadian Telecommunications: An Overview of the Canadian Telecommunications Carriage Industry* (Ottawa: Supply and Services, 1983).

Consultative Committee on the Implications of Telecommunications for Canadian Sovereignty, *Telecommunications and Canada,* Department of Communications (Ottawa: Supply and Services, March 1979).

Department of Communications, *The Canadian Space Programme: Five Year Plan (80/81–84/85)* (Ottawa: Supply and Services, 1980).

Financial Statistics of Canadian Telecommunications Common Carriers (Ottawa: Department of Communications, annual).

ITU, *The Changing Telecommunication Environment: Policy Considerations for the Members of the ITU*, Report of the Advisory Group on Telecommunication Policy (Geneva: ITU, 1989).

C. Christopher Johnston, *The Canadian Radio-television and Telecommunications Commission: A Study of Administrative Procedures in the CRTC*, Prepared for the Law Reform Commission of Canada (Ottawa: Supply and Services, 1980).

Robin Mansell, *Telecommunication/Network-based Services: Policy Implications*, ICCP Report No. 18 (Paris: OECD, 1989).

W. H. Melody, "Implications of US Competition," in *Telecom 2000: Canada's Telecommunications Future*, eds. T. L. McPhail and B. M. McPhail (Calgary: University of Calgary, 1986), pp. 57–76.

W. H. Melody, "Telecommunication: Policy Directions for the Technology and Information Services," in *Oxford Surveys in Information Technology*, Vol. 3 (Oxford: Oxford University Press, 1986).

W. H. Melody, "Telecommunication: Implications for the Structure of Development." *Proceedings, Symposium on Economic & Financial Issues of Telecommunications* (Geneva: 5th ITU World Telecommunications Forum, October 1987).

W. H. Melody, "Telecommunications," in *The Canadian Encyclopaedia* (Edmonton: Hertig Publishers, 1988), pp. 2121–23.

Jean-Pierre Mongeau, Federal-Provincial Examination of *Telecommunications Pricing and the Universal Availability of Affordable Telephone Service* (Ottawa: Supply and Services, 1981).

OECD, *Performance Indicators for Public Telecommunications Operators*, Report by Working Party on Telecommunication and Information Services Policies (Paris: Paris, 1990).

Dallas W. Smythe, "The Relevance of the United States Legislative-Regulatory Experience to the Canadian Telecommunications Situation," Study prepared for the Telecommission (Ottawa: Information Canada, 1971).

Telecom Decision CRTC 85-19, Interexchange Competition and Related Issues. August 29, 1985.

Alvin von Auw, *Heritage and Destiny: Reflections on the Bell System in Transition* (New York: Praeger, 1983).

R. Brian Woodrow and Kenneth B. Woodside, "Players, Stakes and Politics in the Future of Telecommunications Policy and Regulation in Canada," in *Telecommunications Policy and Regulation*, ed. W. T. Stanbury (Montreal: Institute for Research on Public Policy, 1986).

7

Telecommunications Policy for an Information-Intensive Australia

D. McL. Lamberton

AUSTRALIA AS AN INFORMATION SOCIETY

A significant report of the House of Representatives Standing Committee for Long Term Strategies (the Barry Jones report), dated May 1991, underlines the fact that Australia is a highly information-intensive society.[1] Australia is also a small, relatively high-income economy located in a large land area approximately equal to that of the continental United States, with European migration origins and current trade and investment patterns increasingly linked to the Pacific Basin.

This Barry Jones report on Australia as an information society is an appropriate occasion for reconsidering the many questions raised by proposals for a national information policy, not least those relating to telecommunications. In line with its subtitle, "Grasping New Paradigms," the report emphasizes that:

Governments must grasp the significance of the growth of information.

Australia has fallen behind other nations in exploiting information resources.

Australia must recognize the centrality of information as an organizing principle, a tool for understanding, and a vital element in trade expansion.

There is a pressing need to increase the community's use of information.

Coordination of government roles in information activities is needed.

Information issues must be put firmly on the national agenda in the same way that environmental issues have been.

More specifically, the report presses the need for information policy and industry policy to be increasingly integrated. Government should ensure that information-related aspects of industry policies are explored and should encourage industry to become an active user of existing information in research institutions and data bases.

The Committee's principles, propositions, recommendations, and conclusions are grouped under a long list of headings, including telecommunications. Telecom Australia's view that an integrated, coordinated approach to national objectives of an information society was needed was endorsed, as was that body's concern that national communications policies needed to be coordinated with other government policies. At the same time, the report stressed the importance of adhering to the highest international standards for telecommunications.

Two aspects remained in need of clarification. The first is that information appears in national accounts as the cost of making the economic system work as well as being a resource that can, in combination with other resources, play a vital role in development. The spatial and informational dimensions interact; "the tyranny of distance" has been and will remain important in shaping economic development and the role of telecommunications in that process. The second aspect is the derived nature of the demand for telecommunications. Contrary to the impressions given in much of the telecommunications policy debate, investment in telecommunications is intended to provide information services to consumers, business, and government.[2]

AUSTRALIAN POLICY

Under the Telecommunications Act, 1975, the Australian Telecommunications Commission (Telecom Australia) had responsibility for planning, establishing, maintaining, and operating telecommunications services throughout Australia. International services were provided by the Overseas Telecommunications Commission (Australia) (OTC), and Aussat Pty. Ltd., jointly owned by the Commonwealth Government and Telecom, was responsible for domestic satellite communications services.

Telecom had a legislative monopoly over the terrestrial domestic common carrier services; it determined the conditions and prices for private network interconnection with the pubic switched network. Aussat, likewise, had a monopoly over domestic satellite services (but was 25 percent owned by Telecom). OTC had an effective but not statutory monopoly over the provision of international satellite and cable services.

Those responsible for the operation of these services took pride in an evaluation in the International Institute of Communications Phase Two Report: "Despite ... hurdles, Australia has managed to develop a superb domestic and international telecommunications infrastructure, with one of the highest telephone densities in the world."[3] Against this might be set an evaluation by historians under the title, "Service at Any Cost."[4] It is quite impossible in a short chapter to resolve this difference of opinion. One would need to begin with Ann Moyal's fascinating history, *Clear Across Australia*,[5] and wade through the generations of inquiry reports. Even then, doubt would remain because Australia has experienced and continues, to a large extent, to experience policy without research.

THE DEBATE

In July 1987, yet another definitive investigation of the telecommunications system was announced. A national telecommunications task force was set up to examine four broad policy issues: the appropriate nature and extent of monopoly power; the extent of private sector involvement; the desirable extent of restructuring and deregulation of the existing carriers; and how the industry should be regulated, both economically and technically. The objective was a national policy but with emphasis on domestic rather than international aspects. It was expected that the new policy would favor more competition.

The task force deliberations took place in a context of growth and modernization of telecommunications equipment, general information industry development, international competition — with a CoCom cloud on the horizon, foreign investment in the Australian information industry, continuing general controversy on both regulation and privatization, protectionist strategies, and formulation of an official information industries strategy.

The basic issues remained those identified when the task force was announced:[6] more investment in telecommunications; a modernized system, the shift from being technology-driven to demand-driven; segmentation versus integration; the role of the public network; the type of regulatory framework; and arrangements for a smooth transition to the new order. There were, however, some developments that widened the range of considerations, for example, on the international operations front, or limited the scope for policy, for example, fluctuating attitudes against privatization.

Some events that modified the expected course of policy formation related to the following.

INTERNATIONAL OPERATIONS

In March 1988, OTC launched its wholly owned offshoot, OTC International Ltd., following amendment of the Overseas Telecommunication Act, 1987. OTC could design and install telecommunications systems in other countries. Although operation and management of networks was to be limited (for example, Pacific island nations), OTC envisaged a wider market for entire systems, consulting services, and software in the Asian region (for example, Kiribati, Thailand, Sri Lanka, and India) and became involved in the efforts of 15 Pacific island nations to develop a regional small dish satellite network.

Telecom Australia likewise moved to develop international activities in the Pacific region, and there were continuing rumors that Telecom Australia might aim for a merger with OTC and/or Aussat. This new global role for Telecom Australia was boosted by reports that Aussat was in financial difficulties — reports denied by Aussat.

OTC and Telecom Australia initiatives in international operations fit with other developments, although some were hazy and uncertain. For example, the Australian Bond Corporation Holdings Ltd. acquired a 30 percent to 45 percent stake in Companía de Teléfonos de Chile; optimists in the space communications community saw a Cape York spaceport as the Pacific's hub in the future, and others saw the joint Australian-Japanese "multifunction-polis" as the ultimate network.

Whether expectations would be fulfilled was initially problematic. Even the generally optimistic information industries strategy document of the Australian government saw the Australian competitive advantage as limited to design and manufacture of thin route equipment.[7] Those expectations, however, altered the policy context in which the task force deliberated: the international dimensions became much more important.

THE TIMED LOCAL CALLS EPISODE

Telecom Australia had planned to implement its Integrated Services Digital Network (ISDN) system by the end of 1988, and it planned to charge on a time basis. It saw this as the occasion for removing the many tariff anomalies that had developed over many years. However, time charging was linked to a by-election defeat, and the Prime Minister decreed that Telecom's proposal be rejected. In so doing, he noted that Telecom was concerned that significant issues that had led to the particular proposal still existed. The introduction of ISDN was disrupted, and Telecom Australia was left with the problem of devising a politically acceptable tariff system. Not surprisingly, the timing of local calls has remained an issue.[8]

ADVERSE EFFECTS OF DEREGULATION

A report, *Deregulation of the Australian Telecommunications Industry: An Economic Assessment*, by the National Institute for Economic and Industry Research (NIEIR) was commissioned by Telecom Australia, communications equipment industry unions, and the Australian Department of Industry, Technology, and Commerce. It reported on a modeling to 1996 of regulatory effects using data drawn from 38 companies, Telecom Australia, and the unions.

The main findings were that customer premises equipment (CPE) deregulation would cause job losses in Telecom Australia and a net loss of some 5,000 jobs in the economy, temporarily cut gross domestic product by more than $60 million and manufacturing output by more than $80 million (in 1979–80 prices), boost the trade deficit in the short to medium term unless a local content scheme was enforced, increase local phone service charges, and reduce potential development in the local components industry.

It would likely cause a flood of equipment imports, increase the current account deficit, and cut the local equipment industry's output by 7 percent unless existing local content rules were strictly enforced. Removing Telecom Australia's monopoly on first phones would put up to 10,000 jobs at risk (depending on whether Telecom lines terminated outside or inside the home or office) and possibly affect 600 jobs at AWA and STC, two leading equipment suppliers. Deregulation of small business systems (SBS) could see local producers lose from 50 percent to 100 percent of the market to imports and initially boost imports by between $20 million and $40 million annually (depending on local content requirements), although SBS prices would fall by 15 percent to 25 percent. However, most of these losses could be offset by tariff rebalancing, such as a 20 percent to 30 percent rise in local call prices, and increasing local content.

Gains attributed to CPE deregulation included lower prices to business; greater business efficiency; greater availability of cheap handsets to households, providing savings of up to 25 percent; greater choice of first phones; cheaper maintenance and service for some customers (excluding, perhaps, rural and remote customers); and more jobs with private suppliers of services and equipment (at Telecom's expense).

The NIEIR report tempered enthusiasm for deregulation, if only for some, in suggesting that transition problems needed to be taken more seriously. It coincided with the renewal of debate as to the desirability of a general reduction of import tariffs, a traditionally divisive issue in Australia, and possibly reinforced such responses. One significant outcome was the perception of an unacceptable trade-off between small, deferred gains from domestic regulation and the need for strength to compete at the international level.

MAY 1988 DECISIONS

The new policy announced in May 1988 adopted the following objectives:

- to ensure universal access to standard telephone services throughout Australia on an equitable basis and at affordable prices, in recognition of the social importance of these services;
- to maximize the efficiency of the publicly owned telecommunications enterprises — Telecom, OTC, Aussat — in meeting their objective, including fulfillment of specific community service obligations and the generation of appropriate returns on investment;
- to ensure the highest possible levels of accountability and responsiveness to customer and community needs on the part of the telecommunications enterprises;
- to provide the capacity to achieve optimal rates of expansion and modernization of the telecommunications system, including the introduction of new and diverse services;
- to enable all elements of the Australian telecommunications industry (manufacturing, services, information provision) to participate effectively in the rapidly growing Australian and world telecommunications markets; and
- to promote the development of other sectors of the economy through the commercial provision of a full range of modern telecommunications services at the lowest possible prices.[9]

Creation of a regulatory body, the Australian Telecommunications Authority (AUSTEL), raised high hopes of significant gains from a new era of competition in some quarters; others stated bluntly that this was no help in the prospect for the consumer. Although the dust would not settle for some time, some of the changes and some of the remaining issues seemed clear enough. The government had sought to give the appearance of taking back responsibility for communication policy. AUSTEL's role could be of great importance, but primary considerations were its funding and powers. The boundaries between competition and monopoly were shifting. The privatization issue was basically a political one. The interrelationships between OTC-Telecom-Aussat had yet to be worked out.

The process remained a case of policy without research, the NIEIR study being one inception, subject to confidence in the modeling approach employed There was some recognition of research needs. For example, yet another review was announced: the Department of Transport and Communications was to review the ownership arrangements and structural relationship among Telecom, OTC, and Aussat in their conduct of their respective reserved services. This review was to examine, among other issues:

1. The costs and benefits of structural change to the relationship among the carriers, including:
   a. the scope for and costs and benefits of merging one or more of the carriers into a single organization
   b. the desirability of maintaining the present cross ownership and cross board membership arrangements among the carriers
   c. the capital structure of Aussat and options for meeting its financial needs
   d. consideration of whether satellite services should continue to be delivered by single-technology enterprises
2. The future requirements for satellite services in Australia, including the extent to which Australia's satellite service requirements could be met by other satellite systems, and the scope for and the desirability of modifying Aussat's second-generation satellite either by rationalization or by altering its footprint to allow a wider geographical reach.
3. The appropriateness of the boundaries of the respective carriers' monopolies under the Telecommunications Act 1989, including the desirability of regulatory and legislative change to allow greater competition between the carriers domestically and internationally.

The review was to take into account developments in broadcasting policy having a bearing on the role of the carriers, as well as any government decisions in relation to their current references before AUSTEL.

THE NEW "COMPETITIVE" REGIME

The new regime for Australian telecommunications emerged as part of a general microeconomic reform effort and was announced as including the following steps:

the establishment by the end of 1991 of a private sector competitor to a merged Telecom/OTC in the provision of network facilities;

the introduction as soon as practicable during 1991 of competition in the provision of services through introduction of full resale of domestic and international telecommunications capacity (in other words, the current distinction between value added services and reserved services will be eliminated);

three Public Mobile Telephone Services (PMTS) licenses will be issued, including one to Aussat prior to its sale and one to Telecom/OTC, and Telecom's MobileNet will be required to sell airtime on its existing analog system to other PMTS licensees;

the provision of Public Access Cordless Telephone Services (PACTS) will be opened to full competition in accordance with the recent AUSTEL recommendations;

the issues of further mobile (PMTS) licenses will be considered in 1995; and

after June 30, 1997, the duopoly will end.[10]

The key element was reliance on network competition. In line with the earlier shift of emphasis toward concern over international markets, a strong national carrier, the merged Telecom/OTC, was created and provision was made for a second, privately owned network operator to ensure greater innovation and efficiency. Telecom/OTC would be fully publicly owned, while Aussat could compete in providing national and international infrastructure (with a third competitor planned for PMTS).

Because of the critical nature of interconnection arrangements, Telecom/OTC was required to interconnect the new carrier to its network, AUSTEL having power to arbitrate to promote fair and effective competition. The intention was that the second carrier would cover all Telecom/OTC's actual additional costs in providing access to the usage of its network (including allowance for any additional assets required to achieve interconnection and for the opportunity cost of capital).

AUSTEL was given the task of reporting on these matters. In its June 1991 report, two clear principles were drawn from the policy framework: the interconnection charging arrangements should promote greater efficiency levels within a vigorously competitive telecommunications industry, and charging arrangements should promote the development of an openly competitive market as of 1997.[11] These principles led to the following subprinciples:

All charges (both interconnection and usage related) should be set with regard to economic efficiency.

Charges should strike a balance between encouraging a desirable level of facility investment by the carriers and discouraging wasteful duplication of resources.

Charging arrangements should be capable of being applied on a reciprocal basis.

To the greatest extent practicable, the charging structure should be clear and unbundled so that a carrier pays for what it uses and is not forced to pay for what it does not need.

In the interests of administrative efficiency, incentives should be created so that the information requirements of the charg-ing arrangements and of the regulatory regime flow as a natural outcome of the carriers' management information systems.[12]

Usage charges, AUSTEL argued, should be struck so the new carrier would pay fair and reasonable fees on the basis of directly attributable incremental costs. This was interpreted as compensating for the new entrant's lack of market power with a view to establishing charges that would prevail in a strongly competitive industry.

AUSTEL's powers were strengthened, particularly in arbitrating when carriers could not reach a commercial agreement. It has a specific mandate to promote competition and will continue to work closely with the Trade Practices Commission.

RESEARCH AGENDA

Such reviews take on a crisis character. It is, therefore, desirable to consider some of the items that might well appear on the research agenda of AUSTEL and the other institutions equipped to conduct such inquiries.

Sources of Pressure for Deregulations

Should the various reviews and the changes that may well stem from them be seen as a consequence of U.S. export of telecommunications deregulation?[13]

Recent events suggest that budgetary considerations may well have been more important than a basic belief in the superiority of the competitive approach. In addition, there has been awareness of the role of technological change. In the words of the Minister for Communication, "The technology means we simply cannot regulate in the way we used to do, even if we wanted to. We can only have regulations which make sense if they are firmly based in the technology with which we deal."[14] Of course, some appreciate that the theoretical base of the claims for competition needs to be bolstered by faith[15] and that faith is not really a substitute for a detailed mapping of the Australian information sector, with all its complex linkages and multipliers, that would enable consideration of where more competition might prove advantageous.

Information Sector Studies

Far too much policy literature refers in blanket terms to the telecommunications industry or sector. Following the tradition now established by information sector studies,[16] the telecommunications industry needs to be mapped and measured.[17] Given the role of a large unit like Telecom/OTC, this analysis needs to be intraorganizational as well as interorganizational. Without such background knowledge, it would seem difficult, if not impossible, to study effectively the processes of technological change and innovation or to estimate demand for telecommunications products and services, both old and new.[18] The real case for such research is that the demand for telecommunications equipment is a derived demand — a demand that derives from the demand for information. Those who use glib phrases about information being the key to economic survival should reflect on this.

Job Losses

The job loss aspect needs clarification. In Australia, airlines have emerged as institutions in which there was strong pent-up pressure. The 1990 reports, for example, from California, of significant reductions in telecommunications work forces (Pacific Telesis group was reported to be reducing its work force by 11,000 jobs or 16 percent over the next five years[19]) are mirrored in recent Australian developments, although it is difficult to separate the effects of microeconomic changes from those stemming from technological change and regulatory arrangements.

Equity Considerations

As the Barry Jones report has emphasized, access to information and the capability of using information have important equity implications. Given the growth of the information economy and the consequential increased importance of telecommunications, deregulation of telecommunications probably makes information less accessible by lower-income groups at the same time that other changes in technology, for example, computers, may be lowering their relative capability of using information. This suggests that asymmetric information should be high on the research agenda.

Technological Change

Contestability notions have yet to incorporate an adequate treatment of technological change. Thomas Marschak pointed out that "It may well be that only the observing and modelling of real organizations — something economists have been reluctant to do — can led to models of technology and cost which fit usefully into a unified theory of organizational design."[20] It would seem that contestability cannot be separated from technology and organization and the processes by which they change.

Selection of New Technologies

In the Australian situation, a strong case can be made for research into the processes of selection of new technologies. The official Information Industries Strategy is a case in point, as is the Industry Research and Development Board's Generic Technologies Program. One might hazard a guess that in such cases, the selection, both by industry and for policy purposes, was dictated largely by three influences: first, a sensing of great technological possibilities; second, the following of trends perceived in major industrialized countries; and, third, simple projection on the basis of existing interests and past experience. The outcomes from such a selection process may contribute little to Australian competitiveness, because they are reached more on the basis of hunch than on assessment of costs, profitability, and trade potential. A consensus view is no protection if each participant, working scientist/engineer, policy maker, or academic, is subject to the same set of broad influences. Consensus is, therefore, no substitute for a thoroughgoing economic evaluation of technological priorities.

CONCLUSION

The nature of the new "competitive" regime is clouded with uncertainty. It is reported that two consortia are bidding for AUSSAT: Bell South-Cable & Wireless and Bell Atlantic-Ameritch-Hutchinson.[21] Fears are being expressed that the proposed interconnection charges will prove unduly beneficial to foreign carriers.[22] There is a suggestion that two states, New South Wales and Queensland, might turn their networks over to a third private operator.[23] Aussat has initiated a competitive trans-Tasman service with New Zealand,[24] and there is rumor that Aussat's worth might be enhanced by defense links. With so much happening in the open season that began on July 1, 1992, it would seem unwise to make predictions. However, Australian industries of most kinds have remained highly concentrated. In light of competitive experience in other countries, it would not be altogether surprising if this proved true of telecommunications.

NOTES

1. House of Representatives Standing Committee for Long Term Strategies, *Australia as an Information Society: Grasping New Paradigms* (Canberra: Australian Government Publishing Service, 1991). See also D. Lamberton, "The Australian Information Economy: A Sectorial Analysis," in *Challenges and Change: Australia's Information Society*, ed. T. Barr (Melbourne: Oxford University Press, 1987), pp. 13–29.

2. For international linkages see J. V. Langdale, *Transborder Data Flow and International Trade in Electronic Information Services: An Australian Perspective* (Canberra: Australian Government Publishing Service, 1985).

3. Quoted by Senator Gareth Evans (Minister for Communications), "Global Competition and Natural Priorities" (Sydney: International Institute of Communications Conference, 1987), p. 3.

4. N. G. Butlin, A. Barnard, and J. J. Pincus, *Government and Capitalism: Public and Private Choice in Twentieth Century Australia* (Sydney: Allen & Unwin, 1982).

5. Ann Moyal, *Clear Across Australia: A History of Telecommunications* (Melbourne: Thomas Nelson, 1984).

6. See, for example, Henry Ergas, "Telecommunications 2000" (Sydney: International Institute of Communications Conference, 1987).

7. "An Information Industries Strategy," *Australian Technology Magazine*, Special Edition (September 1987): 27–29.

8. "Telecom Needs to Charge for Time Survey?" *The Age*, (April 5, 1991): 5.

9. Statement by the Minister for Transport and Communication, *Australian Telecommunication Services: A New Framework*, May 25, 1988 (Canberra: Australian Government Publishing Service, 1988), p. 3.

10. Australian Department of Transport and Communications (DOTAC), *Micro Economic Reform: Progress Telecommunications* (Canberra: DOTAC, November 1990).

11. AUSTEL, *AUSTEL Study of Arrangements and Charges for Interconnection and Equal Access: Economic and Commercial Considerations* (Melbourne: AUSTEL, June 1991), p. 22.

12. Ibid.

13. Cf. Craig Johnson, "Exporting Telecommunications Deregulation," *Transnational Data and Communications Report* 3 (March 1987): 7–8.

14. Evans, "Global Competition," p. 7.

15. R. R. Nelson, "Assessing Private Enterprise: An Exegesis of Tangled Doctrine," *Bell Journal of Economics* 12 (Spring 1981): 94.

16. See literature from F. Machlup, *The Production and Distribution of Knowledge in the United States* (Princeton, NJ: Princeton University Press, 1962) to M. Jussawalla, D. M. Lamberton, and N. D. Karunaratne, eds. *The Cost of Thinking: Information Economics of Ten Pacific Countries* (Norwood, NJ: Ablex, 1988).

17. R. T. Wigand, "The Communication Industry in Economic Integration: The Case of West Germany," *Social Networks* 4 (1982): 47–79 illustrates some ways of extending traditional industry structure analysis.

18. See "Growing Management Inefficiency at AT&T," *CWA Information Industry Report* (March 1991): 1–3.

19. "Pactel Will Reduce Work Force by 16%," New York *Times*, January 5, 1990, p. C3.

20. T. Marschak, "Organization Design," in *Handbook of Mathematical Economics*, Vol. III, eds. K. J. Arrow and M. D. Intriligator (Amsterdam: North-Holland), p. 1437. It follows that organizational designs should have a place in the

conceptualization of regulation. See D. M. Lamberton, "Australian Regulatory Policy," in *Marketplace for Telecommunications*, ed. Marcellus S. Snow (White Plains, NY: Longman, 1986), pp. 248–49.

21. *Financial Review* (July 2, 1991): 4. Comsat is reported to have sought an alliance with one or other of these. See "Wild Card in Second-Licence Bids," *The Australian* (July 22, 1991): 22.

22. "The Folly of Prices that Profit Foreign Carriers," *The Age* (July 4, 1991): 13.

23. "Report Foresees Emergence of a Third Carrier to Spoil Federal Plans," *The Age* (July 9, 1991): 35.

24. "Aussat's NZ Phone Link Breaks OTC Grip," *The Australian* (June 24, 1991): 20; "Aussat in Great Shape for Impending Sale," *The Australian* (July 13–14, 1991): 4.

8

Privatization, Deregulation, and Beyond: Trends in Telecommunications in Some Latin American Countries

Raimundo Beca

This chapter examines, in a summarized form, the main trends in Latin America in privatization and deregulation issues in telecommunications.

A large use is made of background information collected in the framework of an Economic Commission for Latin American Countries (ECLAC) task force on information technologies, which is preparing a Green Paper in this field. However, nothing is advanced herein about the rationale and the proposals of this Green Paper.

First, the privatization processes in telecommunications in Chile, Argentina, and Mexico are briefly described. At this stage, the rebounding asymmetries among these three processes are underlined. Specifically, it is highlighted that although in Chile two quasimonopolies fragmented vertically (one for local services and the other for national and international long-distance services) were privatized, in Argentina, an integrated monopoly was split into two regional companies (north and south) before its transfer, and in Mexico, the monopoly was sold without dividing it. However, it is also pointed out that to some extent, countries have been learning from each other.

Next, the cases of Brazil and Venezuela, two newcomers still in the privatization pipeline, are analyzed. It is underscored that almost the same dilemmas appear again but receive different responses. The three prior experiences seem to be a helpful prism to appraise these new processes in progress.

CHILE, ARGENTINA, AND MEXICO: THREE DIFFERENT PROCESSES AND ONLY ONE GOAL — PRIVATIZATION

Even though privatization processes in these three countries do have some common traits, including the sale of public concessions and the involvement of foreign telecommunications operators, they are essentially very different. In fact, not only does the degree of private and foreign control differ from country to country, but also the market has been fragmented quite diversely in each of them.

As will be described further, the countries seem to have been learning from each other. A chronological presentation has consequently been retained.

Historical Background

Chile has an old tradition of private management in a multi-operator environment of its telecommunications services.[1] Actually, public involvement in telecommunications covers only the last quarter of a long history that began in 1879, when the first telephone line was installed in Valparaiso, just three years after Bell's invention.[2]

In fact, it was only in the early 1960s, when Empresa Nacional de Telecomunicaciones (ENTEL) was created, that the state became involved in the sector. This commitment was, mainly, the result of its willingness to provide the Chilean economy with an adequate infrastructure for its national and international communications. An earthquake in 1960 raised great political concern about the country's vulnerability in this regard. Specifically, ENTEL was charged with laying out a microwave network trunk, as well as the national and international satellite linkages. Consistently, Chile became the first Latin American member of the International Telecommunications Satellite (INTELSAT), and ENTEL was its signatory.

Public commitment was reinforced in 1971, when Companía de Teléfonos de Chile (CTC) was nationalized during Allende's government. CTC was in those days an International Telephone and Telegraph (ITT) affiliate company whose concessions for local services and long-distance telecommunications (using its own facilities) covered more than 90 percent of the country. Surprisingly, CTC was not denationalized at the beginning of Pinochet's military regime, when a first wave of privatizations was conducted.[3]

Actually, it was only in the late 1980s that ENTEL, CTC, and almost all public services were privatized. In the meantime, the relationship between both companies (traditionally controversial, considering that in practice CTC is ENTEL's only customer) were eased through political arbitrage. At one moment, the government even castled the general managers of both companies.

Notwithstanding, even if CTC and ENTEL held more than 95 percent of the Chilean telecommunications market, competition remained the sector's greatest challenge. In fact, these two enterprises pursued their traditional commercial dispute, although they were publicly owned, but they were also partially exposed to external competition.

Long-distance services theoretically may be provided by CTC or the other local companies either using their own facilities or through third parties. In practice, the market is largely dominated by ENTEL, whose overall shares are 81 percent of income and 75 percent of traffic.[4] The rest is held almost absolutely by CTC. However, two new companies — VTR and Chile-SAT — which recently have been granted concessions to provide long-distance communications through satellite linkages, appear to be quite aggressive.[5]

Regarding cellular mobile telephones, CTC and an independent company, CIDCOM, have been granted overlapping concessions in the metropolitan area and the V region. In the rest of the country, the two bands are shared by two consortia, led by ENTEL and VTR.

Concerning other services, including data communications and leased lines, competition is even more widely open.

Legal Framework

As mentioned before, the present legal structure of the telecommunications sector is the result of the 1982 and 1987 enforcement that shaped the standing General Law on Telecommunications ("Ley general de telecommunicaciones").[6]

This law classifies telecommunications services in four different categories, establishing a particular provision regime for each of them. Additionally, it specifies procedures for granting authorizations and rules for the settlement of tariffs under monopolistic conditions.

Classification of Services

Public services. These are services provided to meet the needs of the community in general (Art. 3-b). The provision of local public services falls under a concession regime. Even though overlapping concessions are theoretically allowed, in practice, holders previously established — CTC, Telefónica del Sur, and TELECOY — benefit from an irreversible dominant position.

Holders of local concessions are compelled after a two-year period to provide services of appropriate quality to all parties requiring it within a compulsory area specified in the grant (Art. 24). Conversely, a certain legal void exists regarding the obligation to provide long-distance and international services.[7]

Intermediate services. These are services provided by third parties to meet the needs of public and limited telecommunications services (Art. 3-e).

Conceptually, this category concerns the facility to provide the service in general and not the service itself. Therefore, the only customers of intermediate services are providers of other telecommunications services.

To launch their appropriate facilities, providers of intermediate services have to request specific concessions, which also have an indefinite extension.

Even though the original intention was to apply this type of concession to services provided by ENTEL, this company succeeded in transforming its old grants into public services concessions. Actually, this category has had for the moment practically no impact at all and has become, as will be discussed further, a source of frequent controversies.

Limited services. These are services provided to meet specific telecommunications needs of enterprises and other independent users (Art. 30-c).[8] These services are not allowed to be linked to public telecommunications networks. Users of limited services have to request a permit for each particular application, and this has to be agreed upon prior to the provision of the service. Permits have a fixed renewable term of ten years.

Complementary services. These are services provided by connecting appropriate equipment to the public telecommunications network (Art. 8). These services may be provided either by holders of public telecommunications concessions or by specific users of their facilities. This concept corresponds in essence to the concept of value added or enhanced services applied in other countries.

Authorization Granting Procedures

With significant differences among services, the standing Chilean telecommunications legal framework accords some discretionary powers to the administrative authority. Conversely, prior legislation in this field assigned only technical attributions to regulatory bodies.

The granting procedures are structured around two basic principles: to preserve overall connectivity of telecommunications services and to avoid unfair competition.

Tariffs Settlement Rules

According to Chilean regulations, tariffs are settled only under monopolistic conditions. Specifically, a tariff is fixed only if two conditions are fulfilled: First, it has to be a public telecommunications service, either local, long-distance, or international. This requirement excludes services such as cellular mobile telephones, leased lines, and intermediate services. Second, the specific market has to be qualified as unfair by the Resolutory Commission ("Comisión Resolutiva"), a special court within the Chilean fair trade enforcement procedure.[9]

Once these conditions are met, the tariff settlement device prescribes that the companies concerned have to present a justified price proposal to the Undersecretary of Telecommunications for each specific service. These tariff proposals have to be based on incremental development costs covering regular operation and investments related to a qualified expansion program and calculated with an appropriate capital rate. Both capital and operation costs have to be comparable to those of an efficient firm in the same field. The capital rate results from the application of a capital-asset pricing model (CAPM), which adds to the market prime rate a systematic risk prime.[10]

If after negotiation there is still a controversy between the enterprise concerned and the Undersecretary of Telecommunications, the opinion of a commission of three experts may be requested. One expert is nominated by each side and the third by common agreement. However, the final decision is taken by the Undersecretary.[11]

Privatization

Analysts currently distinguish, as mentioned, two waves of privatizations in Chile: one in the early days of the military regime, the other as a part of the structural adjustment policies applied in the mid-1980s.[12] During the first phase, with the intentional purpose of obtaining better prices, it was preferred to sell controlling packages or even the whole enterprise. The results were high capital concentration and overindebtedness of the principal economic groups, which later was strongly criticized.

Conversely, learning from the shortcomings of the previous wave, privatization during the second phase was realized following a strategy of progressive sales of noncontrolling packages. However, as some observers have pointed out, the risk of indirect concentration, mainly through financial institutions, is not excluded.[13]

In order to avoid concentration, two new financial sources were widely used. On one hand, the regulatory regime of the pension funds (AFP) was remanaged for this purpose, allowing investment in equities. On the other hand, public workers were authorized and invited to use their long-term service indemnities to buy stocks. Special loans also were accorded to public workers, which, through the pledge of shares, contributed to increasing the control capacity of the banks involved.

In this context, the government's sale of its shares in the two major telecommunications enterprises — CTC and ENTEL — is the result of the conjunction of a wide privatization process, specific entrepreneurial culture, and a new legal framework. Nevertheless, both processes are quite dissimilar: although ENTEL was sold following basically the new privatization model, CTC's transfer was achieved mainly through the transfer of a controlling package.[14]

Like most of the public enterprises during this period, ENTEL was transferred through progressive sales offered preferably to banks, investment funds, pension funds, and public workers.[15] The only exception to this practice was the sale through an international open bid of 10 percent of the voting shares, with an option for an additional 10 percent. Telefónica de España acquired this 20 percent of ENTEL's voting stock, which, given a deconcentration enforcement prescribed by ENTEL's by laws, is the maximum equity position that can be owned by a single holder.[16]

ENTEL's high profit is due mainly to its factual and legal monopoly on international telecommunications, which in 1990 generated $87 million of income, representing 60 percent of ENTEL's revenues.[17] As in other countries, this situation is challenged by both North American carriers which try to set the lower international rates, and national competitors, which plead to enter the market. However, ENTEL has successfully preserved this and other traditional concessions, which were basically obtained in the late 1960s when the Chilean government was willing to grant privileges to its only public enterprise in telecommunications. Nevertheless, after privatization, a certain instability begins to appear at this regard, in that its concessions increasingly are contested by other operators.

Conversely, even if shares were also sold to public workers and pension funds, an open international bid was called in August 1987, offering the control of CTC.[18] According to the specifications of this tender, the bidder initially would get 151 million shares, corresponding to 30 percent of CTC's total stock, and should commit itself to subscribe to its proportional interest in a capital increase of 387 million shares plus those pursuant to the government's preemptive rights in order to complete 45 percent of total stock.

In January 1988 the Chilean government, in a contested decision, awarded this bid to the Australian Group Bond Corporation Holdings. The transfer price was $115 million, which corresponds to $0.76 for each share.

With slight differences to the bid's terms of reference, in the following months, Bond subscribed 215 million additional shares at the same price of $0.76 per share. The total price paid for the acquisition was then $278 million, which theoretically corresponds to 43 percent of total stocks. However, 116 million shares were not subscribed by other holders, which raised Bond's participation to 49 percent. Therefore, the control package was sold at a reference price of $507 per line for the 598,000 lines in service,[19] which, weighted by the 49 percent participation, is equivalent to $1,035 per line for the total capital stock.

Nevertheless, to conform with Chilean general limitations on ownership (Decree-Law 3500), according to which no shareholder may own more than 45 percent of total capital stock, in October 1988, Bond Corporation agreed to reduce its participation below this level by October 1992 (Deconcentration Agreement). In compliance with this agreement, if such deconcentration is not fulfilled by October 1992, CTC has the power to sell such number of shares of the company's stock as to meet the 45 percent limit.[20]

During Bond's administration, CTC began an ambitious expansion program, the target of which was to install 1 million new main lines through 1996, at an annual rate growth of 14 percent. This program, which doubles the Chilean basic telecommunications network, increasing telephone density in main lines per 100 inhabitants from 5.35 in 1989 to 11.75 in 1996, requires a global investment of more than $1.5 million.

In accordance with this expansion program, CTC negotiated a tariff reform that began on January 1, 1989. This new tariff structure, in conjunction with the high rentability of the company, accorded sufficient financial leverage to CTC to obtain suppliers' credit as well as to place debt securities to fund its investment.

In April 1990, Telefónica de España acquired Bond's 49 percent equity position for approximately $392 million.[21] Given the 678,000 lines in service at that moment,[22] this transfer corresponds to a reference price of $578 per line for the controlling package, which, weighted by the 49 percent participation, is equivalent to $1,180 per line for the total capital stock.

Telefónica not only continued and actually implemented Bond's expansion program but also proceeded rapidly to limit its global share participation in advance compliance with the Deconcentration Agreement. With this purpose, 100 million shares were sold in July 1990 through the placement of American Deposit Receipts (ADRs) at the New York Stock Exchange.[23]

Notwithstanding, such legal reform would be insufficient to surmount what happens to be the touchstone of the Chilean deregulation and privatization process: the failure to settle the multiple controversies between ENTEL and CTC.[24] Moreover, experience in this and other countries shows that it is extremely difficult, if not impossible, to legislate in such a highly controversial environment.[25]

The origin of these quarrels is twofold: On one hand is the fact that, strictly speaking, regulations applied to these companies were mainly not those provided in the body of the standing legal framework but those stated in the transitory articles; in practice, this meant that the traditional concessions were frozen even if they were to some extent incompatible with the new enforcement. On the other hand is the privatization of both public enterprises without previous establishment of an adequate mechanism for the prompt settlement of their present or future disputes. In practice, the market alone has been unable to settle these controversies, which were, therefore, referred to courts.[26]

Actually, a series of trials are now in courts, which not only creates an unstable climate but also threatens to paralyze the sector completely. Among these trials, two antimonopoly proceedings dominate the scene: one concerning the provision of long-distance services and the other regarding the joint ownership by Telefónica of CTC and ENTEL.[27]

Long-distance services. This lawsuit had its origin in a request in June 1989 by the Undersecretary of Telecommunications to the Fair Trade Enforcement Office (Fiscalía Económica), raising two specific questions: if fair competition would be affected by vertical integration of the telecommunications sector (that is, the provision of long-distance services by local holders and vice versa) and whether the local and long-distance services should be considered as different markets and whether local companies should be prevented from participating in the long-distance sector. This request was the result of the lack of authority of the Undersecretary of Telecommunications to decide on several grant applications, including, among others, concessions applied by CTC-Transmisiones Regionales S.A., a CTC subsidiary, to provide domestic long-distance services through satellite links and fiberoptic cable systems and a demand by Global Telecommunications S.A., an ENTEL subsidiary, to provide local public services in specific business sectors of Santiago within CTC's concession area. As was mentioned before, it may be argued that the Undersecretary of Telecommunications has the right to decide on these matters according to the provision of article 16 of the law that specifies that a concession can be rejected by this regulatory body, invoking technical or economic reasons. However, after attempting unsuccessfully to mediate a negotiated settlement between both companies, the Undersecretary preferred, at that time, to require the opinion of the Fair Trade Enforcement Office.

In conformity with Chilean fair trade enforcement procedures, a first ruling was issued in October 1989 by the Preventive Commission, which is essentially adverse to CTC.[28] Specifically, it stated that local companies could not participate in the long-distance market and, conversely, long-distance companies could not enter the local market.

Joint ownership by Telefónica of CTC and ENTEL. This lawsuit had its origin in a petition issued in February 1990 by the Fair Trade Enforcement Office to Telefónica because of its significant equity position in ENTEL, so that a consultation about the lawfulness of this transfer should be brought before the Preventive Commission prior to its acquisition of Bond Corporation shares in CTC. Subsequently, in March 1990, Telefónica requested this Commission to state that this joint ownership does not infringe on fair trade regulations. Basically, it was argued that Chilean case law on these issues has constantly reasserted its position that it is not ownership but conduct observed in practice that is the determinant of fair trade. Additionally, it was noted that even considering Banco Santander's shares, the overall equity position is insufficient to control ENTEL; therefore, this would not be a situation of vertical integration.

In April 1990 the Preventive Commission issued a first ruling, stating that Telefónica could not maintain an equity interest in both companies, invoking a potential risk that CTC, given its national monopoly on the local market, may discriminate against ENTEL's competitors. The Commission explicitly took into account the circumstance that not only the behavior of the economic agents is relevant, as Telefónica argued, but also the structure of the market and the regulatory capacity of the administrative authority to affect unfair conduct. Considering that its role is to prevent fair trade infringement, the Commission decided that Telefónica should dispose completely of its equity interest in one of the two companies.

ARGENTINA: THE SALE OF TWO INTEGRATED MARKETS

Historical Background

As in other Latin American countries, telecommunications was introduced quite early in Argentina. Related to the main suppliers of telecommunications equipment, the first operating companies were mostly foreign firms, functioning under a "precarious authorization" regime.[29]

As in Mexico, the prewar years were characterized by strong concentration around two poles: Compañía Unión Telefónica del Río de La Plata, an ITT subsidiary that held almost 90 percent of the market, and Compañía Argentina de Teléfonos (CAT), an Ericsson subsidiary that held 5 percent of the market.[30] Additionally, some small companies were operating services in small communities.

Public involvement with telecommunications operation started in 1946 when the state acquired Unión Telefónica, which became

ENTEL, a wholly owned state enterprise.[31] However, opposite to the strategy followed by Mexico in those days, the new company never merged with CAT, which today still provides services in 6 of the 22 provinces.

A specific feature of Argentina's development in telecommunications, compared with other Latin American countries, is the high weight that the industrial issues have had in the definition of policies. To a large extent, industrial perspectives have predominated over telecommunication's users interests.

In this context, an expansion plan called Megatel was issued, which had the goal of installation of 1 million new lines in five years. To fund this program, ENTEL was authorized to sell telephones in advance with an engagement to install the new lines within 60 months. An equipment decision was finally taken in 1987 under the enormous pressure of 500,000 presold lines. The result was, on one hand, a renegotiation with the traditional providers, which in the meantime were reduced to Nippon Electronic Corporation (NEC) and Equitel. On the other hand, as a part of technical cooperation programs with France and Italy, ENTEL's equipment market was significantly opened to other European providers, mainly Alcatel and Telletrra. Again, national integration was not the principal decisional factor. To the same extent, public telecommunications equipment procurement failed to be the support of the protected electronic national cluster forecast by the informatic policy.

Through the past 35 years, ENTEL had been essentially a passive actor, almost completely in the hands of equipment suppliers. Argentina's economy was suffering dramatically from both scarcity of telecommunication lines and extremely low quality of services. The conditions in Argentina were described as a density of 8.6 main lines per 100 inhabitants, ranking low in the international arena; 47,000 employees for 3.3 million lines, corresponding to a ratio of 14.2 employees per 1,000 lines, which is more than twice the international standard of 6 employees; quality standards of efficiency of calls of 49 percent for local and 29 percent for long-distance, which compares badly with the internationally accepted standard of 99 percent (artificial conditions); an average wait for repair of 14 days, as compared with less than 24 hours in comparable countries; and an installation backlog of more than four years, which is economically unacceptable for a country like Argentina.[32] Additionally, telecommunications services were imposed with a 30 percent tax collected to fund the national pension system.

At this stage, the government concluded that ENTEL's technical, financial, and management problems could be overcome only by significant involvement of a foreign telecommunications operator. In this perspective, an endeavor letter was signed in March 1988 with Telefónica de España, according to which this company would acquire a significant participation in ENTEL's capital and would be charged with its management.[33] Nevertheless, under the pressure of the Peronist opposition, which strongly criticized the lack of competitive bidding, Alfonsín's administration did not finally achieve this privatization.[34]

However, this same opposition, which later won the presidential elections, agreed with the principle of ENTEL's privatization. Not surprisingly, Argentina's President Carlos Menem announced in September 1989 that this objective would be rapidly implemented.

This process concluded in November 1990 with the sale of ENTEL, previously split into two integrated regions, to two international consortia: one with Telefónica as operator and the other operated by Italian and French companies.

Legal Framework

The law states that all telecommunications facilities opened to public correspondence must conform to the National Telecommunications System. International telephonic traffic has to be forwarded through the National Telecommunications System, the only exception being transborder communications, in which case special links could be specifically authorized. Institutionally, the Secretary and the Undersecretary of Communications, under the Department of Public Work and Services, were in charge of telecommunications regulations and the granting of permits and authorizations.

Certainly, this legal framework is ambiguous and unadapted to the present situation. As is discussed later, one of the main shortcomings of Argentina's privatization process results from the fact that this framework was not reformed prior to the ownership transfer.

Filing to some extent this legal gap, specific legislation was passed to privatize ENTEL: the Executive Decree 731/89 of September 1989, which is an implementing regulation of State Reform Law 23696 enacted also in September of the same year.[35] The State Reform Law provides in its main features that it appertains to Congress to decide which public enterprise may be privatized (Art. 9) (a tentative list, including ENTEL, is suggested in an annex); privatizations may be implemented by a sale of assets or shares, as well as by means of granting of licenses or concessions (Art. 17); in principle, the private party is to be selected through international open bids (Art. 18); partial payment in Argentina's public debt may be accepted.[36] Decree 731/89, amended by several successive decrees, specifies the legal framework for ENTEL's privatization.[37] Among other provisions, it authorized exclusion of the transferred companies from the application of the regulatory scheme concerning telecommunications services in order to grant them their respective licenses.

The terms and conditions of the international bid were stated by Decree 62/90 of January 1990.[38] This Decree, called Pliego, defined the structure of the future regime, whose main features would be:

The country is divided into two regions — north and south — of almost equal importance, with the Great Buenos Aires area split between them (Art. 8.2).[39]

The personnel and the assets of ENTEL pertaining to the domestic public area are transferred to two new operators, South Telecom and North Telecom (Art. 7.2).

The personnel and assets of ENTEL pertaining to the international service are transferred to a third corporation (SPSI), whose shares are allocated by halves to the two telecommunications operators (Art. 9.1).

The personnel and the assets pertaining to ENTEL's activities in competition are transferred to a fourth corporation (SSEC), whose shares are also allocated by halves to both operators (Art. 9.11).

The privatization sale concerns 60 percent of the assets of each telecommunications operator (Art. 1.4). The remaining 40 percent will be sold later according to the following distribution: 10 percent to workers, 25 percent to the general public, and 5 percent to cooperatives.

Subsequently, and in parallel with the privatization procedure, a new regulatory agency — the Comisión Nacional de Telecomunicaciones — was created in June 1990 by Decree 1185.[40] This norm redefines the regulatory schemes prescribed by Law 19798, distinguishing now (Art. 21):

licenses, which are necessary to provide telecommunications services (the authorizations and permits in force have to be adapted to the license regime in compliance with a specific regulation to be enacted by the Commission);

authorizations, which are required to install and operate telecommunications facilities according to conditions determined by the Commission;

precarious permits with fixed terms, which are granted to provide temporary telecommunications services whenever it needed to coordinate the entry into activities of several operators; and

precarious authorizations with fixed terms, which are granted to install or operate facilities also during a transitional coordinating period.

Decree 2322 began, nevertheless, in the strongly controversial device of tariff settlement, specifying a new readjustment procedure.

Actually, after a tough negotiation with the bidders, who relinquished a 96 percent rate adjustment they were arguing, they accepted a new formula including a dollar component with a 40 percent weight in the cost of living compensation in case of an exceptional gap between domestic prices and dollar rating, called "adjustments induced by extraordinary and unforeseen events." In return, the licensees agreed to withdraw their right to a 16 percent rate of return recognized by the Pliego. This is, as discussed further, one of the most critical issues of the Argentinean privatization process.

Finally, it should be noticed that not only is Law 19798, with all its ambiguities, still standing, but also new regulations have a hierarchical juridical order, according to which it has to be understood that the Transfer Agreement is submitted to the Pliego and the Pliego itself is submitted to State Reform Law 23696.

Privatization

Perhaps one of the main features of ENTEL's privatization is its origin from a single government officer: Maria Julia Alsogaray. Not only did she have a strong personality, but also she was politically related to a conservative party traditionally opposed to President Menem's justice party.

The split of ENTEL first in four regional companies and finally in two seems to be largely inspired by AT&T's divestiture in several regional operators.

The critical issue of the tender was supposed to be the evaluation of ENTEL's assets, which would serve as a reference for the bidders. Estimated initially to be $3.2 billion (slightly below the company's book value), it was finally stated at a value of $1.7 billion.[41] However, it appeared rapidly that actually, the critical factor for the bidders was the appraisal of the present value at a "reasonable rate of return" and the tariff for the two-minutes reference pulse. Considering annual profits sufficient to fund the investment needed, evaluated at over $500 million per year, a $1.9 billion present value was estimated for a ten-year period.[42]

A two-minute call rate was finally fixed at $0.042, more than twice the highest rate it ever had been in the past.[43] Actually, it could have been 96 percent higher if the bidders had not relinquished the rate adjustment mentioned before.

The two new companies, named Telecom Argentina in the North Region and Telefónica de Argentina in the South Region, took the control of ENTEL on November 8, 1990, only one year after the announcement of its privatization. In any regard, this is a very rapid process for such a big company.

Shortcomings

It is certainly premature to draw up an objective evaluation of ENTEL's privatization. However, it has become such a polemical issue in Argentina that it would be unfair to state that it is not yet possible to establish a first appraisal.

Actually, four main shortcomings are essentially underlined compared with the transfer's objectives mentioned before:

1. Users concern, mainly regarding tariffs, considered excessively high for a service that has not really improved in quality. In fact, the present tariff of $0.042 for a two-minute call is still underrated compared with the $0.05 per minute for local calls estimated with the help of econometric models.[44] However, the main complaints seem to be that this reference tariff for the pulse applied to the monthly basic line access gives a rate of $24 with 200 free pulses.[45] Moreover, Argentinean users were used to extremely low tariffs for a poor-quality service. The improvements in quality standards prescribed by the transfer specifications are not only modest but also are not easily requestable.[46]

2. A network expansion that is considered insufficient to meet the needs of Argentina's economy. Specifically, the installation of 1.2 million new lines through 1996 is requested as an obligatory target, corresponding to an annual growth rate of 5.6 percent, which increases density per 100 inhabitants from 9.8 percent to 12.7 percent. An additional target of 400,000 lines is to be met to obtain a three-year extension of expansion corresponding to a global annual growth rate of 7 percent and an end term density of 13.7 percent. Compared with 1 million new lines installed in Chile by CTC, representing a 13.7 percent rate of annual growth voluntarily decided and the 12 percent of annual expansion prescribed in Telemex's transfer in Mexico, this is a rather modest objective. Also, not only is the additional effort required to obtain an extra period of expansion considered too low, but also it seems in practice quite difficult, if not impossible, to urge the compliance of the bidding objectives. In fact, the only penalty specified by the Pliego (Art. 13.10) is caducity of the exclusiveness in case of repetitive nonfulfillment of the obligatory objections.[47] The experience in many countries that have prescribed such provisions proves that they are extremely inefficient, as expressed by the typical officer reaction: "It's like nuclear weapons, you just cannot use them." Not surprisingly, many observers believe that these objectives probably will not be reached; however, exclusiveness will not be deleted, at least during the first seven-year period.

3. The procurement of public payment funds is also considered insufficient by many observers. However, it should be recalled that ENTEL's final transfer price was 18 percent higher than its current

value, 35 percent higher than its assets value, 35 percent lower than its book value, almost the same as in Chile and Mexico for the controlling package but about half of it for the whole company in these two countries. Certainly the present value after some years will provide an objective answer to this debate. For the time being, it continues to be a highly polemical issue.

4. The establishment of a competitive environment for value added services is generally considered, too, as one of the main short-comings. Actually, the newly created Comisión Nacional de Telecomunicaciones is still in the process of installation. Additionally, value added services have to be defined unequivocally; the notions of "limited services" and "special services," in force, are ambiguous, and they do not cover public value added services. Therefore, firms potentially interested in entering the value added business cannot, in practice, apply for the respective licenses or authorizations.

MEXICO: THE SALE OF AN INTEGRATED MONOPOLY TO MODERNIZE IT

Historical Background

Mexico's telecommunications history is also as old as the telephone, starting in 1878, when a first line linked Mexico to Talplan, only two years after Bell's invention.[48]

As in most Latin American countries, the first years of Mexican telecommunications were characterized by a large proliferation of operating companies, most of them related to foreign enterprises. However, as in Argentina, a strong concentration trend was accomplished before World War II, grouping almost all operating companies around two poles: Empresa de Teléfonos Ericsson and Compañía de Teléfonos Mexicana, affiliated, respectively, with Ericsson and ITT, two main telecommunications equipment suppliers.

In 1948 these two poles merged in a new company, Teléfonos de México (TELMEX).[49] Ten years after TELMEX became owned in majority by Mexican capital, there was a progressive involvement of the state through direct and indirect support to its expansion program, to such an extent that in 1972, the government gained final control over 51 percent of the total shares.[50] The other 49 percent remained in the hands of small shareholders, being publicly traded Mexican Stock Exchange and in the North American market lacer nt of ADRs.[51]

ext, TELMEX achieved a dramatic expansion of unications network from 1.1 million main lines in n in 1988, at a 9.1 percent annual rate of growth.[52]

However, with the economic crisis of the mid-1980s, this dynamism was seriously slowed. Conditions were described as follows: an extremely low telephone density of less than 5 main lines per 100 inhabitants, which compares badly with the 40 lines of Mexico's competitors in the international market; a backlog of more than 1.5 million telephones demanded; a residential penetration of only 18 percent of households; more than 10,000 rural areas with more than 500 inhabitants without access to telephone; a microwave long-distance telecommunications trunk over 20 years old that is highly obsolete and extremely faulty; and a satellite system — Morelos — largely unused because of the overregulations regarding earth stations.[53]

President Salinas de Gortari accorded a high priority to the modernization of Mexico's telecommunications. Moreover, during his electoral campaign he asserted: "Telecommunications will become the corner stone of the program to modernize Mexico's economy."[54] In his view, it would not be possible to attract new investments without an efficient telephone infrastructure and new information services.

In this context, the new Mexican government launched an ambitious modernization program of its telecommunications to privatize TELMEX.[55] It was finally decided to sell, in an international open bid, a controlling package representing 20.4 percent of total shares. The tender was accorded to a consortia formed by the Mexican industrial group Carso and two foreign telecommunications operators, Southern Bell and France Telecomm.

Legal Framework

Even if the new TELMEX's Concession Title introduced some radical modifications to the previous legislation on telecommunications, the ancient regulatory regime is still standing. The main texts, in a hierarchical order, are international conventions signed by Mexico, the 1917 Constitution, the "Ley de Vías Generales de Comunicación y Reglamentos" of 1939, and TELMEX's Concession Title. In its main features, this legal framework classifies telecommunications services, specifies procedures for granting concessions and permits, and accords a preferential regime to TELMEX.

Granting Procedure

With the only exception of the new TELMEX's Title, the regulatory authority has absolute discretional faculty regarding the granting of concessions or permits. The respective grants have a fixed term limit but may be renewed.

Additionally, public services are submitted to tariff regulations, which also have a discretional and not a contractual character. This price regulation device was historically reinforced by specific and extremely high taxes: 90 percent for a local call, 58 percent for long-distance services, and 40 percent for international services.[56] To the extent that the state was reinvesting in telecommunications only one-third of the total amount collected, these taxes were seriously affecting the sector's self-financing capacity.

TELMEX's Preference Concessions

The new TELMEX's Title, agreed precisely in view of its privatization, awards the company a quasimonopoly of Mexican telecommunications structured along the following lines.[57]

1. A monopoly for local services of practically infinite extension. In fact, the old concessions, which had a 30-year term starting in 1976, not only were extended to 50 years but also an automatic renewal for additional periods of 15 years is accorded.

2. A monopoly for long-distance services until at least January 1, 1997. After this date, the administrative authority may compel TELMEX to connect its network to third-party facilities, providing long-distance services so that the user may freely choose the carrier of convenience. In the meantime, TELMEX is not obliged to accept third-party resale of overhead capacity in leased lines.

3. An obligation to expand the basic network in conformity with the following schedule:

Main lines in service should grow at a 12 percent annual rate during the first seven years.

All towns of more than 5,000 inhabitants have to be incorporated into the network before December 31, 1994. From this date on, any telephone installation request should be satisfied within the following six months. This should be reduced to one month by the end of the century.

By the end of 1994, access to telephone — at least through a public coin phone — should be had by all communities of more than 500 inhabitants.

The density of public coin phones should increase, as late as December 31, 1994, from 0.5 phones to two phones per 1,000 inhabitants.

4. A commitment to improve significantly the quality of services, in terms measured by 11 specific parameters, among which are the following:

	Values in 1990	Target for 1994
Lines in fault (%)	10	5
Repaired within 24 hours (%)	45	50
Repaired within 72 hours (%)	82	92
Delay to repair leased lines (days)	5	1
Dial tone before 4 seconds (%)	97	98
Local calls completed — artificial (%)	92	95
Long-distance calls — artificial (%)	90	93
Public phones in service (%)	87	91

5. TELMEX may enter the value added service business. In this case, separate accounting must be kept or activities should be realized by subsidiaries. However, a permit has to be requested from the State Controlling Telecommunication (SCT).

6. TELMEX may also enter the cellular mobile telephone service if a first band previously has been accorded to a third party.

7. A tariff settlement procedure, which accords stability and fairness to price rating. According to this device, a "maximum rate" for a weighted package of services — including connecting charge, basic monthly line charge, and time-measured rates for local, long-distance, and international calls — is settled periodically. A study proposing tariffs for each specific service should be presented every four years by TELMEX. Rates should be calculated considering long-term marginal costs to eliminate cross subsidizing. As in Chile, if there is no agreement between TELMEX and the government on this proposal, the opinion of a group of three experts, nominated one by TELMEX, another by SCT, and the third by a common accord of both parties, might be requested. Whatever, during the first six years, the weighted average price for this package should not increase in real terms. For the next two years, a 3 percent annual reduction in prices should be introduced to transfer to users the benefits of productivity goals. From 1999 on, this factor will be estimated in the cost proposal study.

Privatization

President Salinas de Gortari expressed a strong commitment to TELMEX's privatization and also clearly stated its specific objectives: the state would maintain sovereignty over the country's telecommunications, the rights of the company's workers would be warranted and workers would be invited to participate in its capital, majority control of the company should be retained by Mexicans, quality of services should be improved to meet international levels, and substantial growth of the network has to be accomplished and research and development has to be strengthened.[58]

The international open bid, therefore, was called after an assessing period in which, according to these guidelines, a deal was agreed upon with trade unions, a new concession grant was negotiated with TELMEX, and an evaluation of its assets was charged to international consultants. Contradictory appraisals realized independently by Goldman Sachs, McKinsey, and Booz Allen identified a range from $8 billion to $11 billion for TELMEX's assets, slightly higher than its stock exchange and book values ($8 billion and $7.4 billion, respectively).

Shortcomings

It is certainly too soon to establish a definite balance of TELMEX transfer. However, most observers have been pointing out some shortcoming, such as:

Insufficient level of penetration of the basic network. Even though a compulsory annual rate of growth of 12 percent is settled by the new grant, the present telephone density of 5.2 main lines per 100 inhabitants is so low that in 1996 it will reach only an 8.5 level, far below the 20-line target suggested in ECLAC's recent report.[59] Moreover, given that the only penalty stated by the Concession Title in case of repetitive noncompliance with this obligation is its expiration,[60] there is no guarantee that the expansion objective will be really reached. Actually, as in Argentina's case, the difficulty with this kind of retaliation device is that "you just can't use them."

A tariff and revenue structure inconsistent with the expansion of the basic structure and the suppression of cross subsidies. In fact, even if tariffs were considerably raised in 1990, just before privatization, local revenue per line forecast for 1996 is only $269 because $168 per line is required to fund the 12 percent expansion plan.[61] A 62 percent of operation margin would thus be required if self-funding is pursued.

Even though long-distance services' tariffs were also heavily raised in 1990, shifting form $0.17 to $0.28 per minute, the present rate seems insufficient to fund the basic network expansion.[62] Actually, long-distance tariffs are comparable to North American rates reached after several years of competition.

The permanence of Telecommunications Mexico (TELE-COMM) in the public sector. Even if SCT's traditional operation activities were transferred to TELECOMM and some of them to the private sector, including TELMEX for the microwave long-distance links, many observers feel that President Salinas' will to privatize Mexico's telecommunications has not yet been completely fulfilled. Moreover, TELECOMM's activities are protected by Mexican Constitution, according to which it has the monopoly of telegraph and broadcasting and, by extension, that of data transmission and data processing. In the past, the country has known a certain underdevelopment of these and other value added services. Most likely, as telecommunications will continue to converge with data processing and audiovisual communications, strong pressure in favor of liberalization of this will develop.

A precarious regulatory body. In fact, not only is the SCT insufficiently provided in both budget and human resources, but also, as noticed earlier, the whole legal framework has to be restructured in order to suppress ambiguities and legal voids and to assure its consistence with TELMEX's concession grant. On the positive side, one should definitely include the deep modernization process TELMEX is experiencing since its privatization. Even if some concerns have been pointed out regarding certain shortcomings, there is no doubt that quantitative and qualitative improvements of major importance are in course.

A significant contribution in this direction is the strong commitment of trade unions to the process, specifically to the stamping out of corruption.

Foreign operators' advice in management also is a key factor. However, conversely with ENTEL's case in Argentina, a minimum of outside appointment has been required.[63]

In what, for most observers, has been a big surprise, Grupo Carso has shown in practice its willingness to be an active partner and a real leader of the consortia. Actually, perhaps anticipating potential competitors, Carso launched immediately an aggressive $9.5 billion investment plan for the first three years, approximately 50 percent higher than the program considered in the Concession Title. Moreover, matching both the research and development objectives and the human resources training, a Mexican Institute of Telecommunications was created and provided with an adequate budget.

In sum, even if such obstacles jeopardize the whole process if they are not promptly overcome, its seems that globally, Mexico is achieving successfully its main goal: privatize TELMEX to modernize it.

BRAZIL AND VENEZUELA: TWO NEWCOMERS FACED THE SAME DILEMMAS

TELEBRAS and CANTV, the Brazilian and Venezuelan public monopolies in telecommunications, are now in the privatization pipeline. In both cases, as in Argentina and Mexico, there is strong political commitment in this regard by new governments in a structural adjustment context. Additionally, poor management is generally invoked as one of the main reasons in favor of privatization.

The two companies are vertically integrated, covering local, long-distance, and international services. To this extent a big similarity exists with ENTEL and TELMEX.

TELEBRAS is a holding company covering vertical and horizontal market segments. Chile's and Argentina's market fragmentations are thus extremely relevant, too. However, both kinds of market segmentation also have been present among the different options open to CANTV.

Brazil: An Act Still Waiting to Be Performed

As in the famous Pirandellos' theater piece "Seven Actors Searching for a Theater Author," Brazil's privatization is still finding its way. In spite of the strong commitment of President Collor de Mello and the former Minister of Finance Zelia Cardoso to telecommunication privatization, the Brazilian government has failed to overcome the establishment's rebuttal. Trade unions, senior management, equipment suppliers, members of congress, and other opposers of TELEBRAS's privatization have found in the Brazilian Constitution an extremely powerful ally.

As in Chile, government's involvement in the telecommunications sectors covers only the last quarter of a long history that goes back to the early days of telephone invention, when Graham Bell made a personal gift to Emperor Pedro II.[64] The rationale for public commitment was the lack of dynamism of private companies that were providing telecommunications services, to such an extent that with only 1 million lines for 70 million inhabitants, Brazil had one of the lowest telephone densities in the world in the mid-1960s.

EMBRATEL, a public enterprise, was created in 1965 and not only acquired Companhia Telefónica Brasileria (CTB; a Canadian subsidiary), which concentrated approximately 60 percent of the total network but also extended progressively its coverage by the acquisition of a large number of small companies to which local concessions had been granted in the past. In parallel, an investment fund (Fondo Nacional de Telecomunicaciones) was created, to which a 30 percent tax on telecommunications was allocated.

The Brazilian national telecommunications system was completed in 1972 with the creation of TELEBRAS, a public holding company, structured around three main principles: 24 state telecommunications companies (one per state except for Rio Grande do Sul, where the local company is held directly by the state) in charge of services within their territory; EMBRATEL in charge of interstate and international communications as well as special services, including data communications; CPqD, a research and development center.

The federal government's monopoly and, thus, TELEBRAS' exclusive rights were granted not only by the Telecommunications Code, enacted in 1962, but also by its inscription in Brazil's 1988 Constitution, which specifically states in its article 21 that it is the Union's competence to "exploit directly or through concessions awarded to firms under public control, the telephone, telegraph, data communications and other public telecommunications services." Additionally, EMBRATEL was granted by a regulatory act ("portaría") an extension of its monopoly on data transmission to private communications.[65]

In the first years after telecommunication nationalization, TELEBRAS grew at very rapid rates. The Fondo Nacional de Telecomunicaciones was the main source of financial resources. However, with the crisis of the mid-1980s, this dynamism disappeared almost completely.

Finally, the current situation of TELEBRAS, increasingly criticized by sectors in favor of its privatization, presents four main weaknesses: an extremely low penetration of telecommunications services, as characterized by a density of only six main telephone lines per 100 inhabitants and regional inequities in a ratio of 1 to 15;[66] a poor management efficiency, the main consequences of which are low quality standards, such as 11 employees per 1,000 lines, 39 percent calls completed, 84 percent of dial tone after three seconds, 9.2 percent of lines installed but not in service, and 5 percent of telephones out of service;[67] exceedingly high investment costs, estimated at over $3,000 per line,[68] about twice the price range of $1,300–$2,200 observed in other Latin American countries (partly because of the protection of local suppliers and the lack of competitiveness of Tropico's switching equipment,[69] a product developed by CPqD, this high investment cost represents an enormous handicap for the network expansion); a tariff and cross-subsidizing structure that is arbitrary and precarious. Arbitrariness results from an atypical chain of cross subsidies characterized by the facts that 50 percent of the revenues of local companies comes from long-distance services and that, conversely, data communications represent 35 percent of EMBRATEL's income.[70] In other terms, data communications are subsidizing long-distance services and long-distance services are subsidizing local services. Precariousness results form the fact that both EMBRATEL's data transmission monopoly and its exceedingly high prices are increasingly being contested by its main users, Brazilian and foreign business.

Venezuela: Privatization of an Integrated Monopoly to Modernize It, Revisited

Challenged also by low density and poor quality of services, Venezuela's government is undertaking CANTV's privatization with what some opponents feel is an excessively accelerated pace.[71]

CANTV was originally a private company founded in 1930 and granted a nonexclusive concession the same year.[72] The new company rapidly acquired the numerous firms that had been granted concessions since the first line was installed in Caracas in 1883,[73] only seven years after Bell's invention.

Over the years, CANTV was progressively nationalized in a process that culminated in 1968, when all the shares became publicly owned. In parallel the company's concession shifted to a contract in compliance with Venezuela's telecommunications legal framework and, mainly, with the country's Constitution. Actually, as in Mexico and Brazil, Venezuela's Constitution accords to the Union the monopoly of telecommunications, stating in its article 136 that all matters concerned with them are of the competence of the National Power.[74] However, this enforcement does not go as far as in Brazil, where concessions can be granted only to public enterprises. Specifically, the standing legal framework prescribes that the establishment and operation of all telecommunications systems by all means known or invented in the future "correspond" to the State (Art. 1 of the Law of Communications)[75] and present telecommunications services or others that in the future the government would decide to add will be "centralized" in CANTV (Art. 3 of the Law that reorganized telecommunications services).[76] Subsequently, a contract was signed in 1965 charging CANTV with providing the following services:[77] local, long-distance, and international services; national and international telex services; facsimile; telephoto; data transmission; facilities to transmit radio and telecommunications programs; and provision of telegraph channels.

As mentioned before, CANTV's development is heavily criticized. Three main concerns are usually invoked:

Excessively low penetration of basic network. Even if its density of 9.1 lines per 100 inhabitants is one of the highest in this region, this value is strongly moderated with respect to Venezuela's gross domestic product.[78]

Extremely poor quality of services. The following areas, in fact, were mentioned in a prospectus calling for investment in CANTV's privatization:[79] 52 percent unsatisfied demands, 49 percent of completed calls for local services, 31 percent of completion of long-distance calls, 24 percent of international calls completed, 11 workers per 1,000 main lines.

Unacceptable levels of corruption. These are recognized even by opponents to CANTV's privatization, who state that the company is not profitable due to bad management and corruption.[80] It is specifically charged that political "clientelism" has been largely exercised in hiring and that political parties, public agencies, and even some private enterprises are not paying their bills.

In this context, Venezuela's government called for a prequalification of international telecommunications operating companies. At that time, what was offered to operators was a threefold package:[81]

1. An option to purchase 30 percent or more of CANTV's shares, to be exercised at the operator's convenience. The rationale behind this formula was to grant a certain financial relief to investors in order to fund the modernization program.

2. A concession to provide telecommunications services, which in fact would regularize a certain ambiguity concerning the lawfulness of the transfer of CANTV's right.

Finally, ten international operators were retained, including NTT, France Telecomm, Bell Canada, GTE, and several regional Bell operating companies.

Considering suggestions by some operators and the relative success of a mobile cellular telephone license bid conducted recently, in which offers were much higher than expected, the government decided to modify the privatization scheme, calling for sale of a package encompassing the acquisition of 40 percent of CANTV's shares; the direct management of the company, not only its lease; and a nine-year exclusive concession. Additionally, quality standards and expansion objectives are imposed on the holder.

In parallel, the government has been carrying out the study of a new legal framework, for which some support has been received from international agencies. Some observers have, nevertheless, expressed their concern regarding the fact that both processes — privatization and regulation — were being conducted by independent teams with little interaction between them. Fortunately, a certain convergence has been reached recently.

In sum, it should be noticed that Venezuela has taken some clear options on telecommunication privatization dilemmas, including maintain integration, regulate after privatization, impose quantitative and qualitative objectives, sell only a controlling package, grant only mid-term exclusiveness (nine years), and transfer the management to a foreign operator. It is difficult to know to what extent these decisions were inspired by other countries' experiences. Whatever, there is no doubt that other countries had some influence on past decisions and that they will also have some influence on dealing with remaining issues such as regulations, a regulatory body, tariffs, and cross subsidizing.

IS THERE A LATIN AMERICAN MODEL FOR TELECOMMUNICATION PRIVATIZATION?

The question about the existence of a Latin American model has often been raised, receiving relatively contrasting answers. From past experiences, we would be tempted to reply that this is not yet the case.

Actually, as is shown in Table 8.1, it may be asserted that there are more differences than likeness between the three privatization processes described above.

This table shows the big asymmetries among these three processes. However, one may also identify a certain number of winning solutions.

TABLE 8.1

Comparison of Privatization Process in Latin American Countries

Issues	Chile[a]	Argentina[b]	Mexico
Prior density[c] (lines per 100 inhabitants	4.9–5.4	9.8	5.2
Initial number of lines (thousands)[d]	548–678	3,175	4,300
Initial quality levels	Local calls completed, 97%; long-distance calls completed, 93%; repair in two days, 75%; lines with fault, 7%	Local calls completed, 49%; long-distance calls completed, 29%; Number of days to repair, 14; lines with fault, 45%	Local calls completed, 92%; long-distance calls completed, 90%; repair in three days, 82%; lines with fault, 10%
Market fragmentation before privatization	Vertical	None	None
Initial management conditions	Adequate	Extremely poor	Poor
Rationale of telecommunication privatization	Not specified	Implicit, including users demand quality network expansion public payment funds competitive environment for value added services improve management	Explicit, including government sovereignty quality rights of its workers Mexican control of the company network expansion research/development

Foreign debt	No	Yes	No
Workers' involvement	No	No	Yes
Package sold	49% of the shares of a local service company	60% of the assets of two regional companies, which share, in halves, value added services and an international services company	20.4% of the shares of an integrated company, representing 100% of voting shares
Market fragmentation after privatization	Vertical	Horizontal	None
Transfer price (million US$)	278–392	1,350	1,758
Equivalent transfer price for the package (US%/line)	507–78	425	390
Equivalent transfer price for the whole company (US$/line)	1,035–348	705	1,915
Nature of the grant	Nonexclusive concession for local services	Exclusive license for local, long-distance, and international services during an initial term, nonexclusive license thereafter	Exclusive concession for local services, exclusive concession for long-distance services for six years
Time extension of concessions	Indefinite	Seven years of exclusivity with a possible extension of three more; infinite thereafter	Fifty years, automatically renewable for fifteen-year periods

Table 8.1, continued

Issues	Chile[a]	Argentina[b]	Mexico
Limitations to foreign participation	None	None	At least 51% of Mexican capital in the 20.4% controlling package
Management agreement	None	Agreed to an experienced telecommunications operator holding at least 10% of the consortia's shares	None
Sector's regulation	Prior	Partially after transfer	After transfer
Efficient regulatory body	No	No	No
Government's presence in the board of directors	No	No	No
Tariff settlement	Only for monopolistic services; every four years for each service, based on incremental cost compared with an efficient firm	Tariffs were settled before and are automatically adjusted in real terms with a productivity reduction factor of 2% per year, which becomes 4% in the additional period	Every four years for a set of services based on incremental cost compared with an efficient firm; the weighted rate for the set has to diminish in real terms by a factor of about 3% per year
Quantitative obligations	None	1.2 million new lines (5.6% per year)	12% per year

Qualitative obligations	None	For 1996 — local calls completed, 85%; long-distance calls completed, 80%; days to repair, 3; lines in fault, 30%	For 1994 — local calls completed, 95%; long-distance calls completed, 93%; repair in 3 days, 92%; lines in fault, 5%
Penalties	Only abolishment	Only abolishment	Only abolishment
Local services market	Open	Open after seven or ten years	Monopoly
Long-distance service market	Dominant position of ENTEL	Open after seven or ten years	Open after six years
International service market	Dominant position of ENTEL	Open after seven or ten years	Open after six years except satellite services
Value added services	Open; the holder may participate	Open; the holder may participate through SSEC	Open; the holder may participate
Mobile cellular service	One band in two regions	May apply for a second band	May apply for a second band
Present local monthly rates (after privatization)	Line charge $9.50; time measured rate $0.028 per minute	Line charge $24 (includes 400 free minutes); additional rate $0.021 per minute	Line charge $3.50 (includes 150 free calls and 200 free minutes); additional call $0.011; additional minute $0.035
Equivalent local monthly rates[d]	$26.30	$28.20	$18.10
Forecasted density (1996)	11.8	12.7	8.5

[a] Where two values are given, the first refers to Bond's acquisition in 1988 and the second to Telefonica's acquisition in 1990.
[b] Argentina's transfer price was estimated considering a 17 percent quotation for its foreign debt papers.
[c] Initial conditions correspond to those prevailing in the respective transfer year.
[d] Considering 200 calls of three minutes per month.

NOTES

1. Narui Raul Dominguez, "El proceso de privatization en el sector telecommunications de Chile." Mimeo, ENTEL, 1990.

2. Ibid.

3. Mario Marcel, "Privatizatión y finanzas públicas: el caso de Chile 1985–89," in *Estudios Cieplan No. 26*, Santiago, Chile (1989).

4. Estimated by Ingetel Company.

5. José Ricardo Melo, "Liberalization and Privatization in Chile," presented to the seminar "Implementing Reforms in the Telecommunications Sector: Lessons from Recent Experiences," Washington, April 1991.

6. "Ley general de telecommunicaciones," No. 18.168.

7. There is a wide bibliography in this respect. A bill has been sent by the Chilean government to Congress that transforms the indefinite extent of broadcast and telecommunications service concessions (which use the radioelectrical spectrum) into, respectively, a 30- or 60-year renewal term ("Proyecto de Ley," Secretaria General de la Presidencia, July 1991).

8. In fact, limited services also include third parties' services to meet the needs of broadcasting operators.

9. The Chilean fair trade enforcement procedure prescribes three jurisdictions: the Preventive Commission acts as a first-ruling court; its decision, which is not binding, may be appealed to the Resolutory Commission; this second ruling may be, in turn, appealed to the Supreme Court.

10. Haroldo Miranda, "La legislación tarifaria chilena: Espíritu, forma y aplicación al caso de servicios de telecommunicaciones regulados." Mimeo, Subtel, 1989.

11. "El colegio de Ingenieros de Chile y la Ley General de Telecommunicaciones," Santiago, 1990.

12. Marcel, "Privatizatión y finanzas Públicas."

13. Ibid.

14. Actually, TELEX CHILE, a company that provides telegraphic and telex services, was also privatized in this period. However, we have preferred, given its comparatively minor significance, not to consider this company in our analysis.

15. ENTEL, *Annual Report, 1990.*

16. In fact, the board of directors didn't agree to modify ENTEL's by laws in order to accept a higher concentration level.

17. ENTEL, *Annual Report, 1990.*

18. Corfo: "Bases para la licitación de 151.000.000 acciones de la CTC y posterior suscripción de aumento de capital de 387.000.000 de acciones," Santiago, Agosto 1987.

19. CTC, "Anuario Estadístico General de Desarrollo Telefónico CTC, 1990."

20. Salomon Brothers, IFC, "American Depositary Shares Representing 100.000.000 Shares of Series A Common Stock," Prospectus, July 1990.

21. Ibid.

22. CTC, "Anuario."

23. Salomon Brothers, IFC, "American Depositary."

24. Luis Terol, "CTC, una compañía privada con orientación de mercado," CTC, May 1991.

25. Melo, "Liberalization."

26. "El colegio de Ingenieros."

27. Salomon Brothers, IFC, "American Depositary."

28. Comisión Preventiva Central, "Dictamen 718/705," October 16, 1989.

29. Héctor A. Mairal, "The Argentinean Telephone Privatization," presented to the seminar "Implanting Reforms in the Telecommunications Sector: Lessons

from Recent Experiences," Washington, April 1991.

30. Alejandra Herrera, "La revolución tecnológica y la telefonía argentina," Legaza, Buenos Aires, 1989.

31. Ibid.

32. Compilation by CEPAL/ONUDI from several sources.

33. Herrera, "La revolución."

34. Mairal, "The Argentinean Telephone."

35. Ibid.

36. "Reforma de Estado," Ley No. 23696, September 12, 1989.

37. "Decreto Plan No. 731/89" (Empresa Nacional de Telecommunicaciones), September 12, 1989.

38. "Pliego de bases y condiciones," January 5, 1990.

39. Actually, the proportions established by the Pliego were 53.3 percent for North Telecom and 46.7 percent for South Telecom. Regarding the Buenos Aires area division, the boundary line was defined as North and South Cordoba Avenue. It should be noted that originally, it was meant to divide ENTEL in four regions. As a matter of fact, Decree 731/89 stated that three regions were to be considered in the international open bid.

40. "Creación Nacional de Telecomunicaciones," Decreto 11185/90, June 22, 1990.

41. Alejandra Herrera and Ben Alfa Petrazzini, "Privatización de los servicios de telecommunicaciones: el caso Argentina," mimeo, June 1991. As these authors note, the book value estimated by the Sindicatura de Empresas Públicas for 1987 was $3.5 billion. The reference transfer price for the bid was fixed by Decree 420/90 at $1,003 million for the 60 percent package, $1,671 million for the entire company.

42. Ibid. As these authors point out, a 30 percent rate of return was used for this estimation, which is extremely high given that cost of living and exchange rates adjustment are warranted.

43. Ibid.

44. Robert W. Crandall, "After the Breakup: U.S. Telecommunications in a More Competitive Era," Brookings, Washington, 1991.

45. Herrera and Petrazzini, "Privatización."

46. "Pliego."

47. Ibid.

48. Instituto Mexicano de Comunicaciones, "Situación actual y perspectivas de las telecomunicaciones en México." Documento preliminar, mimeo, October 1990.

49. Alfredo Pérez de Mendoza, "Teléfonos de México: Development and Perspectives," in *Changing Networks: Mexico's Telecommunications Opportunities*, eds. Peter F. Cowhey, Jonathan D. Aronson, and Gabriel Szekely (San Diego: University of California, 1989).

50. Instituto Mexicano, "Situación."

51. Mendoza, "Teléfonos."

52. Secretaria de Comunicación y Transporte, "Programa de modernización de las telecomunicaciones de México," Mexico, 1990.

53. Calos Mier y Terán, "Modernización de las telecomunicaciones de México," 13th Pacific Telecommunications Conference, Honolulu, January 1991.

54. Quoted by Gabriel Székely from Jesús Sánchez, "La coordinación de comunicaciones, punta de lanza del esfuurzo moderniizador, Salinas," El Financiero, October 1, 1987. See also Gabriel Székely, "Mexico's Challenge: Developing an International Economic Strategy," in *Changing Networks: Mexico's Telecommunications Opportunities*, eds. Peter F. Cowhey, Jonathan D. Aronson, and Gabriel Székely (San Diego: University of California, 1989).

55. Terán, "Modernización." *Changing Networks: Mexico's Telecommunications Opportunities*, eds. Peter F. Cowhey, Jonathan D. Aronson, and Gabriel Székely (San Diego: University of California, 1989).n."

56. Instituto Mexicano, "Situación." *Changing Networks: Mexico's Telecommunications Opportunities*, eds. Peter F. Cowhey, Jonathan D. Aronson, and Gabriel Székely (San Diego: University of California, 1989).n."

57. "Modificación al título de concesión y Reglamentos" of 1939; D. Alfonso Arenal, "Mexico," in *Régimen Jurídico de las telecomunicaciones en Hispanoaméroca*, AHCIET (Madrid: Government Press, 1990).

58. Terán, "Modernización."

59. Raimundo Beca, "Información y telecommunicaciones: Las exigencias de desarrollo Sutentable," Mimeo, CEPAL/ONUDI, July 1990.

60. "Modificación al título."

61. "Diagnóstico sobre niveles de desarrollo y utilización de las technología de la información en Mexico." Division CEPAL/ONUDI, in preparation.

62. TELMEX, "Indicadores importantes de TELLMEX: Supestos en los que se basan las proyecciones," 1990.

63. Carlos Casasuz, "Privatization of Telecommunication: The Case of Mexico," presented to the seminar, "Implementing Reforms in the Telecommunications Sector: Lessons from Recent Experiences," Washington, April 1991.

64. Ministerio de Comunicaciones, *Anuario 1984.*

65. Portaría No. 525 of 1988, which amends prior texts. Strictly speaking this monopoly presently covers only interstate private data communications. However, EMBRATEL benefits by an externality effect.

66. TELEBRAS, "Relatorio Comercial," February 1991.

67. TELEBRAS, *Annual Reports 1987, 1988, 1989, 1990.*

68. Salomon Wajnberg, "Sector de telecomunicaciones: Argentina, Brasil, Chile, Paraguay y Uruguay: Servicios e Industria," Mimeo, BID-INTAL, July 1990; Joao Carlos Fagundez Albernaz, "Telecommunications and Data Services — Brazilian Situation," (Budapest: TIDE 2000, 1990).

69. Claudio Frichtak, "Specialization, Technical Change and Competitiveness of the Brazilian Electronics Industry," OCDE workshop, "Technology's Change and the Electronic Sector: Perspective and Policy Options for Newly Industrialized Countries," Paris, June 1989.

70. TELEBRAS, "Relatorio Comercial," December 1990.

71. Movimiento Profesional, Antonio José de Sucre, "Ante la Privatización y desnacionalización de las telecomunicaciones de Venezuela," Caracas, 1991.

72. José Mora, Vicente López, and Rosita Baroni, "Venezuela," in *Régimen jurídico de las telecomunicaciones en Hispanoamérica*, ACHIET (Madrid, 1990).

73. Ibid.

74. Ibid.

75. "Ley de Comunicaciones," July 12, 1940.

76. "Ley que regula la Reorganización de los Servicios de Telecomunicaciones," July 1991.

77. Mora, López, and Baroni, "Venezuela."

78. Beca, "Información."

79. Venezuela's Investment Round Table, "Opportunities in the Telecommunications Sector," Caracas, March 1991.

80. Sucre, "Ante la Privatización."

81. Venezuela's Investment Round Table, "Opportunities."

9

Telecommunications in Africa: Policy and Management Trends

Raymond U. Akwule

Studies that focus on the problems of telecommunications advancement in developing countries have often concluded that a major hurdle is the poor organization and management of the sector.[1] In many developing regions — including all the African countries — telecommunications equipment and service offerings, as well as rates and conditions, are usually controlled by a government monopoly, but there is growing evidence of shifts in attitudes and practices toward less rigid government domination and more market-sensitive telecommunications regulatory, policy, and management environments in some countries. The shift has both historical and developmental import for Africa and should be of supreme interest to the international telecommunications community.

The goal of this chapter is to review the policies, past and present, especially those related to the organizational and managerial practices, that have guided the conduct of African societies in the provision of telecommunications services. The new organizational and management trends will be illustrated by reviewing the practices in three countries — Nigeria and Senegal in West Africa and Kenya in East Africa.

THE PAST

In the early 1960s, the methods of organization and management of telecommunications in the newly independent nations of Africa bore strong resemblance to those of the European colonial powers that dominated the continent for more than a century. The new government leaders devised national telecommunications policies with little or no consideration for the prevailing market forces. The policies across Africa generally provided for substantial governmental financial subsidy of the sector, and in many cases, governmental control was absolute. National posts and telecommunications were typically administered as part of a unit, and international telecommunications was managed under a separate umbrella.[2]

The 1970s was an important decade of change in attitude toward telecommunications in Africa. During the decade, there was substantial increase in the worldwide recognition of the potential of telecommunications-aided economic and social development.[3] Also during this period, African leaders became increasingly aware that the level of available telecommunications on the continent was inadequate to meet the developmental tasks ahead. In some cases, available telecommunications were concentrated in the urban areas, leaving the rural areas, where the majority of the population (80 percent) resides, without any facilities.

Various collaborative efforts were initiated, aimed at bridging the huge and widening gap in the levels of available telecommunications between the continent and the rest of the world. For example, a plan for a Pan-African Telecommunications (PANAFTEL) network was initiated, with the initial objective of interconnecting the national networks but later with the added objective of enhancing the entire public telecommunications network down to the subscriber level.[4] Regional organizations, such as the Pan-African Telecommunications Union, were created to oversee and foster the integrated development of various aspects of telecommunications on the continent. Also, in a attempt to emphasize the severity of the communications problem in Africa and to direct international attention toward seeking solutions, the United Nations (UN) proclaimed 1978–88 as the United Nations Transport and Communications Decade in Africa (UNTACDA). As part of UNTACDA, a target of 1 percent telephone density in Africa by the year 1988 was set. Furthermore, the International Telecommunications Union, the specialized UN agency charged with responsibility for the development of telecommunications in remote areas of the world, has focused much of its activities on Africa.

However, the various national and international collaborative efforts to solve Africa's communications problems so far have resulted in relatively little progress. For example, the continent did not attain the development goal of 1 percent telephone density by 1988. Furthermore, the pace of growth in communications infrastructure and services has been very uneven across the continent, and even within the nations,[5] and though many obstacles to development of the sector have been identified,[6] observers have often concluded that a major hurdle to communications advancement in Africa is the poor organization and management of the sector.7

THE PRESENT

Since the mid-1980s, several African countries have started a process of restructuring their telecommunications sectors in an effort to meet the challenges of an emerging worldwide telecommunications society that is increasingly proliberalization and proprivatization. The change has been induced by two major factors: increased national and international pressure on the telecommunications administrations to provide better telecommunications services, and a push toward austerity programs and economic reforms by the International Monetary Fund and North American, European, and Japanese banks that hold billions of dollars in potentially unrecoverable debt to many of the countries.

Across the continent there has been a trend toward the following modes of practices: separation of telecommunications and postal operations, merging of national and international telecommunication systems under one management umbrella, less reliance on government subsidy of telecommunications and concurrent increase in nongovernmental investment, and more autonomy for the national telecommunications regimes accompanied by increased reliance on private-sector styles of management in the administration of national telecommunication services.

The degree and mix of changes have differed from one country to the next. For example, Nigeria and Senegal have made significant structural changes in recent years that include separating the administration of the postal and telecommunications sectors, while Kenya has continued to exercise controls. Nonetheless, policies and practices in all three countries reflect a zeal to cope with increasingly complex national and international demands on the telecommunications sectors.

NIGERIA

This West African nation, with a population of more than 100 million, is by far the most populous nation in Africa. A former British colony, Nigeria has in its three decades of national independence experienced both economic prosperity (resulting from an oil boom in the 1970s) and economic depression (tied to the decline in the oil market and the worldwide economic recession in the 1980s). Economic hardship in the 1980s forced the Nigerian federal government to initiate a structural adjustment program that would affect telecommunications among other economic sectors.

Nigeria started moving away from very rigid centralized governmental control of its telecommunications industry in 1985 when, in an effort to improve efficiency in the country's communications system, the government reorganized the Ministry of Communications and the agencies that it oversees.[8] Before the reorganization, domestic telecommunications were the responsibility of the Posts and Telecommunications Department (P&T) of the Ministry, and external telecommunications were administered by the Nigerian External Telecommunications (NET). Reorganization meant separating the domestic telecommunications services that were under the P&T and merging them with the external telecommunications services that were provided under NET, to form the Nigerian Telecommunications Limited (NITEL). NITEL and the Ministry of Communications now share responsibilities for the regulation and administration of telecommunications in the country.

NITEL is a government-owned limited liability company. The company has wide-ranging operational autonomy, and its policies are subject to review and approval by a board of directors.[9] The federal government has placed significant pressure on NITEL to improve the company's efficiency and to attain self-sufficiency, and NITEL has responded by increasingly injecting private-sector-styled management practices in pursuit of those goals. The company has a three-tier administrative system that is composed of the headquarters, the zonal, and the territorial management levels.

The headquarters defines the business policy of the company, sets operational and technical standards, carries out macroplanning, undertakes viability studies for the big projects, and provides directives and guidelines for use nationwide. The zones are operational centers that implement the business policies of the company. They also plan and provide telecommunications services and for the management of revenue collection. The territorial units are responsible, on a day-to-day basis, for the services given the customers and for the prompt collection of revenue.

The territories are expected to break even, and competition is encouraged between the territories and between zones. Each operational unit is assessed by criteria such as revenue generated, revenue actually collected, telephone and telex lines requested in relation to lines installed, time period for correcting faulty lines, and so on.

Two major company objectives have been to revise the country's telecommunications tariff structures upward to reflect current international telecommunications rates and to institute a more effective bill collection program aimed at reducing the huge customer debt owed to NITEL. Accordingly, since 1985, there have been dramatic increases in telecommunications tariffs. For example, in 1988, international telephone tariffs were increased by as much as

700 percent, depending on the type of call.[10] In addition, an aggressive bill collection program has been established.

The reorganization coupled with a national telecommunications tariff revision and the aggressive bill collection program has helped to improve the financial picture of the country's telecommunications sector. As evidence, the Ministry of Communications announced that total revenue generated by NITEL during the first half of 1988 was 111 percent more than the amount generated during the same time period in 1987. The Ministry attributed the increase in revenue to the decentralization of NITEL's billing system.

The telecommunications sector has continued to have its problems, some of which, ironically, are caused by the reorganization of the old P&T and by the new management practices introduced by NITEL. For example, reorganization of the sector has meant reduction in the size of the work force in the old P&T bureaucracy. There has been resentment on the part of those who lost their jobs as a result of the trimming exercise. Furthermore, a combination of dramatic increases in the cost of national and international telephone calls over just a few years and the drastic penalties imposed on telephone subscribers for nonpayment of telephone bills has engendered some anti-NITEL feelings in the citizens. It is reasonable to conclude that from the point of view of the Nigerian public, the shift toward deregulation and private-enterprise-styled telecommunications regimes in Nigeria has come at a very high price.

SENEGAL

Senegal offers another example of the new telecommunications sector organizational and managerial trend in Africa. The changes in the telecommunications sector are part of a wider program to stimulate economic growth in that West African nation. A French colony until 1960, Senegal, which still has strong ties with France, experienced severe economic difficulties in the late 1970s. The economic problems were caused by a combination of some national government policies, some external factors, and recurrent drought. However, the country has taken steps in recent years to arrest the downward economic trend.

Liberalization of the economy has been introduced as part of a national strategy. Quantitative import restrictions have been gradually eliminated; the final phase of tariff reduction took place in July 1988. A timetable for the liquidation, privatization, or rehabilitation of some 60 quasi state organizations has begun. Companies remaining in the government portfolio will have their subsidies reduced and will operate under stricter performance criteria. The new policies are affecting the telecommunications sector.

Until October 1, 1985, telecommunications management in this West African nation was entrusted to two separate bodies: the Office of Posts and Telecommunications (OTP), which was responsible for managing and operating the national network, and the National Society of International Telecommunications of Senegal (TELESENEGAL), which was responsible for the operation of the international network.

In 1985, the government of Senegal dissolved the OTP and set up two new entities: the Office of Posts and Savings Banks and the National Society for Telecommunications (SONATEL). The latter was formed by extending the jurisdiction of the former TELESENEGAL to cover the operation of the national network as well.

SONATEL operates as a national company with the state as the whole shareholder. Its status as a national company allows it the flexibility to operate a totally autonomous "private enterprise" type of management. Alasane N'Diaye, Director General of SONATEL, outlined his country's rationale for reorganizing its telecommunications industry to reflect a more commercial trend. Under the old arrangement, he said, in a statement released in 1985, that Senegal had a highly efficient international network (TELESENEGAL) that was known to be one of the leading companies in the country in terms of financial results and the quality of its services. However, the performance of TELESENEGAL stood in striking contrast to that of the national network (OTP), which had become known for its inadequacies, including an extremely high equipment saturation rate and very antiquated infrastructure, mediocre quality of service, a telephone density of 3.4/1,000 inhabitants, an unmet 50 percent demand, an estimated productivity of 70 technicians/1,000 main lines, and extremely low levels of investment in the sector.

In 1985, because of the existence of obsolete cable networks and the lack of adequate maintenance, approximately 15 percent of the national network was said to be out of service at any given time. In addition, the average domestic call completion rate was less than 50 percent during peak hours. However, the international telecommunications facilities (which consisted of one type A satellite earth station and four submarine cables with terminal switching equipment) were quite well-maintained and, therefore, had low fault rates. Unfortunately, even the international system could not be fully exploited because the local network was so poor.

SONATEL has since introduced dramatic changes in its functions, structure, and operations. Modern management techniques have been introduced with the aim of instilling an atmosphere of healthy competition and enthusiasm among the workers. For example, bonuses and incentives have been increased for all levels of management to reward merit, endeavor, devotion, and commitment. According to N'Diaye, "This is necessary, since the so-called 'private sector' mentality must be instilled and once-and-for-all replace the 'civil servant' mentality." He further stressed that the new management practices are geared toward regaining the confidence of financial backers and ensuring attainment of the objectives set by the government, which include speedy improvement of the quality of service offered to subscribers; completion on time of projects underway; optimum use of available human resources in the company; ultimately, securing easy access to the telephone for every Senegalese; enhancing the prestige of telecommunications; and modernizing the network by gradually introducing digital technologies.

SONATEL may not have met all of its objectives yet, but it has succeeded in dramatically improving the availability and quality of telecommunications service in the country. For example, the number of main telephone lines nearly doubled between 1985, when Senegal had approximately 22,000, and 1990, when there were more than 41,000 main telephone lines.

In addition to rapid expansion, the quality of telecommunications service has improved tremendously. For instance, in 1985 the telephone fault rate averaged one fault per subscriber every three months. In 1988, the rate was one fault per subscriber every eight months. By June 1990, the rate had dropped to an average of one fault per subscriber every 20 months.

The fault correction rate also has improved considerably in recent years. In 1985, less than 70 percent of the faults reported were corrected within one week. By the beginning of 1991, more than 95 percent were corrected either the same day or within seven days following the fault report.

The financial picture of the sector is much better, too. SONATEL's 1990 earnings totaled 985,000,000 CFA francs, as compared with 193,000,000 CFA francs in 1986. In addition, the cash resources rose from 3,586,000,000 CFA francs in 1986 to 11,109,000,000 CFA francs in 1990. One can only predict that the future will be even brighter for Senegal's telecommunications sector if the prudent management of the past several years persists.

Several other African countries have implemented, or plan to implement, changes similar to those described in the examples from Nigeria and Senegal. Yet, in some countries — such as Kenya — development objectives are being pursued without major structural changes in the telecommunications sector.

KENYA

When Kenya attained independence in 1963, its policies were geared toward enhancing that nation's telecommunications network in preparation for the development tasks ahead. For many years,

Kenya participated with Tanzania and Uganda in a customs union/ common market, and the three countries jointly operated a number of important services, including airways, railways, and post and telecommunications. On July 1, 1977, the customs union arrangement collapsed, causing Kenya to create its own national telecommunications administrative agencies. Two agencies were created: the Kenya Posts and Telecommunications Corporation (KPTC), to oversee Kenya's national telecommunications, and the Kenya External Telecommunications Company (KENEXTEL), to manage the country's international network. In 1982, KENEXTEL was merged with KPTC.[11]

A major motivation for telecommunications development in Kenya during the 1980s has been the desire to spur economic activity in the rural areas. There has been national concern over the excessive rural-to-urban population drift in the country, a migration caused by the lure of employment opportunities in government agencies and businesses located in the nation's capital and some other urban centers. According to the World Bank, in 1965, 9 percent of Kenya's population lived in the urban centers, but by 1987, the percentage of the population living in the urban centers has risen to 22 percent.[12] There is apprehension that the rural-to-urban migration might eventually hurt the nation's economy by crippling the agricultural sector, which has its base in the rural areas and accounts for approximately one-third of the gross domestic product. As a result, the Kenyan federal government has initiated a "District Focus for Rural Development" strategy, as part of an overall plan to reduce the rate of rural-to-urban migration.

The KPTC, for its part, has taken steps to slow the rural-to-urban migration by stressing telecommunications establishment and growth of businesses in those areas. "It is the policy of this government to provide the necessary infrastructure that will stimulate economic activity especially in rural areas," stated Minister for Transport and Communications Arthur Magugu on the occasion of the tenth anniversary of the founding of the KPTC in 1987.

In addition to expanding basic telecommunications services to the rural areas, KPTC also is in the process of modernizing its network.[13] To fuel the expansion and modernization programs, KPTC more than doubled the size of its staff, from 5,356 employees in 1980 to 12,000 employees in 1988. Two-thirds of the staff are described as technical personnel.

The increased pace of network expansion and modernization has led to large increases in importation of network equipment. Imports for the telephone and telegraph industry, which include switching and switchboard parts and apparatus, rose from $10,665,198 in 1987 to $13,719,354 in 1988, an increase of 28.6 percent. Continued sector growth is expected in the coming years as a consequence of an ambitious government program to provide a total of 500,000 lines by the end of the year 2005. The program will result in Kenya having a ratio of one telephone line for every 20 people by that year.

The Kenyan federal government prescribes the nation's telecommunications policies and sets broad operational goals through the Ministry of Transport and Communications, which supervises KPTC. KPTC has had a monopoly over all telecommunications and postal matters, including responsibility for the allocation, licensing, management, and control of the radio frequency spectrum and for the management and provision of all telecommunications services. However, policies instituted in 1991 now allow other private companies to market and manufacture new telecommunications equipment after satisfactory completion of the projects. In addition, it is expected that KPTC's influence will remain dominant, because the corporation will continue to be the most important purchaser of communications equipment in Kenya.

According to KPTC's managing director, the liberalization policy is aimed at stimulating participation of Kenyan-based telecommunications companies, which are now allowed to perform services such as internally wiring buildings, providing telecommunications equipment, and procuring, marketing, installing, and providing service backup for various terminal equipment. The decision to liberalize also is said to be part of the government's plan to encourage local participation in the manufacture and assembly of electronic and telecommunications equipment.

A step toward accomplishing that goal was taken in December 1988, when the KPTC established a multipurpose manufacturing plant in the Kenyan city of Gilgil. The project aims to help Kenya achieve self-reliance in telecommunications and to provide equipment for export to east, central, and southern Africa.[14] The plant, which employs approximately 1,000 people, can assemble 100,000 telephones per year. In addition, it can produce cableforms, switchboards, power units, and other telecommunications equipment. The KPTC is especially targeting export of the Gilgil products to other countries within the Eastern and Southern Africa Preferential Trade Area region. Orders already have been received from Tanzania and Uganda. In one of its biggest export ventures, which took place shortly after the Gilgil plant was opened in 1988, the KPTC signed a three-year, multimillion dollar agreement to supply cableforms to African Telecommunications manufacturing plant in Northern Ireland.

KPTC's efforts to provide a telecommunications network that will meet the basic needs of its citizens, including the need for fairly good communication with the international community, seem to have been moderately successful. As already indicated, a central theme of the telecommunications development programs in Kenya has been the zeal to generate economic activities in the rural areas by expansion of communications facilities to those areas. There is evidence that the expansion and modernization programs have resulted in availability of some telecommunications facilities in most of Kenya.

Unlike the other two countries described in this chapter, Kenya continues to combine the posts and telecommunications sectors under the same management entity. Nonetheless, the revolution in the telecommunications sector is no less significant than those of Nigeria and Senegal. Kenya has continued to institute changes — the latest of which include some liberalization of the telecommunications sector — in an effort to meet national developmental objectives.

CONCLUSION

The changes in Africa's telecommunications policies and practices deserve special attention not only because they represent a very radical departure from past practices in the region but also because of the unique socioeconomic context within which telecommunications services are provided in the continent. Disproportionate concerns about poverty in this region relative to other world regions have made telecommunications policy choices more difficult than elsewhere.

Some short-run effects of the continent's new trend toward procompetitive managerial practices in telecommunications sectors are already apparent. One such major short-term effect has been increased revenues in some countries that have adopted the new approach. For example, in Nigeria, NITEL officials have attributed the company's increased profits in recent years to the company's new "business-like attitudes."[15] On the other hand, the new "efficiency syndrome" has resulted in a reduction of the work force in the old bureaucracies, resulting in many disgruntled jobless citizens. In several countries, including Senegal, there have been various forms of social protests directly attributable to the "efficiency measures."

The long-term impact of these measures is difficult to assess. On the one hand, African telecommunications regimes are eager to attract both domestic and foreign (private and public) investment capital to the telecommunications markets and, by extension, stimulate their economies, and the trend toward more efficient, market-oriented telecommunications regimes is generally seen as essential to the attainment of that goal. On the other hand, in Africa, where the countries have been severely afflicted by the problems of recent worldwide economic recession and where drought has taken its toll on much of the population, it should not be surprising if further attempts to deregulate and "commercialize" telecommunications regimes are greeted with strong public resistance and even social unrest. This is one risk that the already-fragile political entities in most of the region are reluctant to take. Perhaps it is the major reason some African telecommunications administrations are slow to embrace the worldwide trend toward deregulation, liberalization, and privatization.

NOTES

1. Robert J. Saunders, Jeremy J. Warford, and Bjorn Wellenius, *Telecommunications and Economic Development* (Baltimore: Johns Hopkins University Press, 1983); Sir Donald Maitland, *The Missing Link: Report of the Independent Commission for Worldwide Telecommunications Development* (Geneva: International Telecommunications Union, December 1984).

2. Many African countries left the operation of their international services in the hands of the same international companies that had managed them in the colonial era, such as the British-owned Cable and Wires Company and the French-owned Societe France Cable et Radio.

3. Heather Hudson, Douglas Goldsmidt, Edwin B. Parker, and Andrew P. Hardy, *The Role of Telecommunications in Socio-Economic Development: A Review of Literature with Guidelines for Further Investigations* (Geneva: International Telecommunications Union, 1979); Saunders, Warford, and Wellenius, *Telecommunications*.

4. A more comprehensive communications development project that goes by the name of Regional African Satellite Communication System for the Development of Africa (RASCOM) has since been initiated. The goal of the RASCOM project is the establishment of efficient and well-managed communications infrastructural facilities for the continent as a whole. The RASCOM project, which emphasizes a prominent role for satellite communications, is intended to supplement the accomplishments of the PANAFTEL project.

5. For example, the North African countries of Egypt, Libya, Algeria, and Morocco have average telephone densities that far exceed those of their neighbors to the south, except South Africa. In general, there has been a bias toward development of communications infrastructure and services in the urban centers.

6. Raymond Akwule, "U.S. Cooperation for Africa," in *Agenda for Action*, edited by Anatoly Gromiko and C. S. Whitaker, pp. 129–35. (Boulder, Colo.: Rienner, 1990).

7. Saunders, Warford, and Wellenius, *Telecommunications*; Maitland, *The Missing Link*.

8. Raymond Akwule, "Telecommunications in Nigeria," *Telecommunications Policy* 15 (June 1991): 241–47.

9. Information about telecommunications in Nigeria was obtained from original documents obtained from the Nigerian Telecommunications Company Ltd., *Seminar on Telecommunications Policy for Nigeria (January 26–February 6, 1987)* (Lagos: Ministry of Communications, Federal Government Printer, 1987) and from various issues of *Africa Telecommunications Report*.

10. The Board of Directors comprises some NITEL directors and some members not affiliated with NITEL.

11. "Nigeria: A Special Report on NITEL and Nigerian Telecoms," *Africa Telecommunications Report* 4 (February 1989): 1–5.

12. Information about Senegal was summarized from original document from Senegal's national telecommunications agency, SONATEL, and from

"Development of Telecommunications in Africa," *Africa Telecommunications Report* 2 (October 1987): 4–7.

13. Kenya's recent announcement of the impending end to the KPTC monopoly over the provision of all telephone services in the country is an example.

14. "NITEL Profits Up," *Africa Telecommunications Report* 3 (October 1988): 2.

REFERENCES

Africa Telecom '86 Book of Speakers' Papers: World Telecommunications Forum. Geneva: International Telecommunications Union, 1987.

Raymond U. Akwule. "African Communications in an Information Age." In *Soviet U.S. Cooperation for Africa: An Agenda for Action.* Boulder: Lynne Rienner Publications, 1990.

Raymond U. Akwule. "Telecommunications in Nigeria." *Telecommunications Policy* 15 (June 1991): 241–47.

Raymond U. Akwule. "Telecommunications Policy and Management Trends." In *World Telecommunications Forum — Africa Telecom '90: Development Strategies for Resources Management and Technology.* Geneva: International Telecommunications Union, 1990.

"Development of Telecommunications in Africa." *Africa Telecommunications Report* 2 (October 1987): 4–7.

Jeff Fishbein. "Kenya KP&TC: Ten Years of Progress." *Africa Telecommunications Report* 2 (September 1987): 2.

Heather E. Hudson, Douglas Goldsmidt, Edwin B. Parker, and Andrew P. Hardy. *The Role of Telecommunications in Socioeconomic Development: A Review of Literature with Guidelines for Further Investigations.* Geneva: International Telecommunications Union, 1979.

Independent Commission for Worldwide Telecommunications Development. *The Missing Link.* Geneva: International Telecommunications Union, 1985.

"Nigeria: A Special Report on NITEL and Nigerian Telecoms." *Africa Telecommunications Report* 4 (February 1989): 1–5.

NITEL Journal 2 (August 1987).

"Kenya: End to KPTC Monopoly in Sight." *Africa Telecommunications Report* 5 (August 1990): 1–3.

"Kenya's PTT Signs Cableform Contract with AT&T." *Africa Telecommunication Report* 4 (February 1989): 1–5.

"NITEL Profits Up." *Africa Telecommunications Report* 3 (October 1988): 2.

Robert J. Saunders, Jeremy Warford, and Bjorn Wellenius. *Telecommunications and Economic Development.* Baltimore: Johns Hopkins University Press, 1983.

Seminar on Telecommunication Policy for Nigeria (January 26–February 6, 1987). Lagos: Ministry of Communications, Federal Government Printer, 1987.

10

Telecommunications Development in China: Problems, Policies, and Prospects

Lin Sun

People's Republic of China ("China" hereafter) is the most populous country in the world, with 75 percent of its 1.2 billion people living and working in the rural areas. China is among the poorest countries; its gross national product (GNP) per capita in 1989 was between $308 and $330 (World Bank 1990; Hauser & Laughlin 1991).

Like most developing countries, China's economy is operated by a central government. Government ministries control the country's political, economic, and social lives, although there has been an increasing trend to relaxing economic control in recent years. The central regime is also the brainchild of socialist ideology, which contends that the social wealth should be shared by all citizens. The ideal, envisioned by Marx and experimented with by Lenin and Mao, seems to have lost its popularity recently, as Eastern Europe has abandoned the socialist system and turned to Western democracy and market economy.

The central government in China also controls the telecommunications industry. Since the early 1980s, telecommunications has been granted a national priority, and its growth has been remarkable. Ironically, the improving economy and technology seem to be driving the industry in a different direction, with growing local autonomy and decentralized financial sources, production, and distribution. The fundamental forces behind this movement are market demand, financing shortages, sector competition, and technology.

FUNDAMENTAL CHARACTERISTICS

Industry Structure

Parallel to its political structure, China's telecommunications industry operates essentially under a centralized regime. On the top of the structure is the Ministry of Posts and Telecommunications (MPT), the regulator and primary equipment and service provider. The MPT is a huge bureaucracy. It consists of 18 departments and committees, including Directorate General of Telecom (DGT), planning, policy and regulation, construction, wireless communications, and postal services. In addition, MPT owns about 30 large and 90 small-to-medium manufacturing facilities and 35 research centers throughout the country. Altogether, the MPT employs about 1 million people and has assets worth at least 50 billion yuan ($45 billion).

The power of MPT is also shown in its operating branches in all provinces and major cities throughout the country. Each MPT department has its provincial offices under the umbrella organization called "telecommunications administration bureau." This central mechanism is to ensure that MPT's policy is executed effectively. For example, DGT controls 30 local DGT branches in all provinces.[1] The provincial DGTs command their affiliates in cities and counties, and so on. Figure 10.1 illustrates the hierarchical structure of China's telecommunications industry (the postal administration has a similar structure but is separated from telecommunications).

The "top-down" administration structure was established right after the founding of the People's Republic in 1949 and has remained ever since. In the early years, such highly centralized control was believed to be the best option when the country was trying to provide very basic service to a large population but had scarce capital and technological resources. Apparently, the central industry regime worked well, especially in the 1950s. Posts and telecommunications business grew 200 percent per year between 1953 and 1957. Telephone density (sets per 100 population) increased from 0.05 in 1949 to 0.13 in 1957, despite population growth of about 3 percent per year during the same period (Liang & Zhu 1988).

Historically, China's centralized structure made a remarkable contribution to the early development. Experiences in other developing countries seem to support this conclusion (Wellenius et al. 1989). However, any industrial policy and structure are very sensitive to changes in market and technology as the industry grows. The legitimacy of a centralized structure should be reevaluated in the face of these changes.

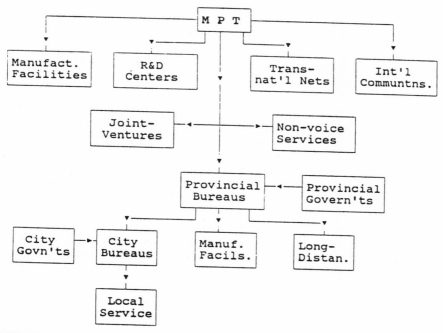

FIGURE 10.1
Structure of Telecommunications Administration in China

State Protected Monopoly

A centrally controlled industry often means a state protected monopoly. It is especially the case in China, because China is also a socialist country that commands the concentration of economic and political power and social wealth. In a socialist state, government is the legitimate candidate for control of resources, production, and distribution. Because telecommunications is often regarded as a public good, government becomes the sole owner and regulator of the industry. Many in China believe a government-controlled industry can yield optimal economies of scale because it is operated in a planned manner, whereas market forces would dilute the scarce resources by unruly competition.

In telecommunications, the incarnation of socialist doctrines is MPT. Like other government agencies, the MPT monopoly is endorsed and protected by the government in the form of legislation. The MPT enjoys all favorable political, financial, and technological treatments. The protection, at least in form, is formidable. For instance, there has been growing sentiment about the efficiency of MPT's monopoly and its capability of meeting the best interest of the market because there is little rivalry pressure for doing so.

Meanwhile, there has been an increasing tendency to reduce the central control of manufacturing, product distribution, and service. The government's response to these challenges is adamant and fierce. In 1989, the State Council in its approval of MPT's structural reform dismissed the idea of diluting MPT's role as the industry regulator and administrator, reiterating that the MPT "is the supreme administration of telecommunications nationwide." The government even granted the MPT "professional and technical management responsibilities for private communications networks" (Zhou 1990). In reality, however, the MPT seems to be sliding away from its vested role as the market continues to grow and diversify.

After the early development, the state protected monopoly has made the telecommunications market inflexible. As the industry has expanded, the MPT has also grown to an industry colossus; its presence makes any meaningful competition nearly impossible. The MPT institutes regulations and technical policies for the market it dominates. As a result, its monopoly is actually reinforced. In a preempted market like this, non-MPT entities are likely to be defeated by low product quality and incompatibility, therefore, they are unlikely to gain significant market shares. Voluntary withdrawals may occur sooner or later, because the profit is thin in such an environment. The MPT's monopoly is also strengthened by its control of the research advantage in introducing new products before anyone else. The following are detailed accounts of MPT's monopoly by major industry sectors.

Manufacturing Facilities

All central office (CO) exchange manufacturing facilities are directly under MPT's supervision. This includes switch circuits, technical specifications, and standard regulations. Given the complexity of manufacturing and technical requirements, CO exchange manufacturing has remained successful in the MPT's territory, and its growth has been quite fast. Table 10.1 shows that the total exchange capacity (urban and rural) in the 1980s had an average annual growth of 9 percent (of which the urban increased about 14 percent). The switching capacity in the urban areas increased much faster than that in rural areas. The gap is even greater if stored-program controlled (SPC) automatic switching is taken into account.

The exchange market consists of digital and analog switches, both dominated by the MPT for the past four decades. The MPT introduced the first digital SPC private branch exchange (PBX) in 1982. By 1989, SPC exchanges had a share of 40 percent of total CO switch production, most of which were manufactured by the MPT. An inspection report issued by the MPT reveals that of 99 time-division multiplexing and space-division multiplexing PBX products, the MPT makes at

TABLE 10.1
Growth of Total Telephone Switching Capacity (1980-89)

Year	Urban Ex. (10,000)	Growth Rate (%)	Rural Ex. (10,000)	Growth Rate (%)
1980	200.3	—	242.9	—
1981	217.9	8.8	245.5	1.1
1982	240.9	10.6	249.8	1.8
1983	262.2	8.8	253.9	1.6
1984	292.0	11.4	261.6	3.0
1985	336.5	15.2	276.9	5.9
1986	380.5	13.1	291.9	5.4
1987	464.5	22.1	309.4	6.0
1988	555.8	19.7	n.a.	—
1989	1,035.0*	—	—	—

*Urban and rural capacity.

Source: Liang, X., & Zhu, Y. (1988, September). The development of telecommunications in China. Paper presented at Conference on Pacific Basin Telecommunications, Tokyo; The use and development of local telephone exchanges. (1989c, November). World Telecommunications, 2, 10–12; China telecommunications services and investment in 1989. (1990, May). World Telecommunications, 3, 55.

least 54, a share of 55 percent (World Telecommunications 1989b). The MPT also dominates the crossbar switch products, most of which are supplied to rural networks.

Other government agencies, notably the Ministries of Electronics, Railways, Petroleum, and Defense, also manufacture and implement about 20 percent to 25 percent of PBXs. These private communications systems are emerging strongly in recent years, and some have even carved into MPT's monopoly by providing quasipublic services such as leasing lines to outside users. These ministries have good research facilities, and their market presence has formed some rivalry to the MPT.

Transmission Networks

Since the early 1950s, the MPT has been the sole supplier of transnational trunk construction and maintenance through its centralized facilities.

The term "network" in China entails two types of coverage: domestic and international. All international voice communications are switched and routed via MPT exchanges and international communications ports in major cities such as Beijing, Shanghai, and Guangzhou (Canton). The domestic network is more complicated, containing thee parts: long-distance, local (urban), and rural.

The long-distance networks consist of five levels: interprovincial, provincial centers, intercounty, county centers, and terminal exchanges (Liang & Zhu 1988). There are six interprovincial centers in the country, located in major traffic hubs. Calls are switched at these national centers to provinces and transmitted forward to call destinations. Such transmission configuration is parallel to the administration hierarchy described earlier. For example, because interprovincial hubs are responsible for crossprovincial traffic, central control from MPT is imposed. Likewise, if traffic is essentially within a province, the local telecommunications bureau naturally assumes responsibility for management and maintenance. In such a hierarchical service structure, competition is hardly viable, given the strict control in distribution.

Terminal and Customer Premises Equipment

This sector is probably most competitive in all the sectors examined. The terminal and customer premises equipment (CPE) market used to be owned and controlled by the MPT. Since the mid-1980s, however, MPT's role has been gradually relaxed. The reason is not MPT's willingness to relinquish its power but the result of market dynamics. First, it has become difficult for the MPT alone to meet the broad range of demands that have increased dramatically. Second, the structural reform since the early 1980s has allowed "horizontal" consolidation among different government agencies (see next section). Third, both the MPT and non-MPT companies are now engaged in technology transfer and joint ventures with Western companies, increasing production and market competitiveness. Last, the technical requirements for most CPE are relatively easy to satisfy as compared with CO exchanges and trunk systems. The outcome is certainly encouraging: terminal equipment suppliers are competing for the captive operating equipment market (OEM), whereas CPE (mainly telephone sets and accessories) is becoming a hybrid and lucrative market.

Enhanced and Nonvoice Services

Although the basic telephone services still have a long way to go to reach the majority of the population, this does not prohibit MPT from engaging in large-scale research and development (R&D) and trials for new services. Since the late 1970s, the MPT has conducted numerous research projects beyond plain old telephone service (POTS), such as videotex, intelligent networks, and system access protocols. In the past two years (1991 and 1992), China has successfully experimented with digital switching with fiberoptic transmission in metropolitan telephone networks. Chinapac, the first public packet switching X.25 network, has been in service, providing a prototype of videotex in Chinese characters, electronic mail, and telephone directory assistance. Trials of integrated services digital networks are also being planned. Because these are essentially centralized services, they are exclusively provided by the MPT. Such initiatives taken by the central government in developing countries is often seen as an advantage rather than a deficiency in harnessing technical and economic resources.

Distinction should be made between local telecommunications authority and the MPT. Although the former is a subsidiary of MPT, it in reality exercises certain autonomy, particularly when local funding and construction are involved. This is the key to understanding competition within a central regime in China. The aspects of local autonomy are discussed in the next section.

Financing

Compared with other infrastructural industries, telecommunications requires high and long-term capital investment. Not only does the constraint lie in the scope and value of the investment, but also the economic return is often indirect and delayed. This latter aspect is the critical barrier to increasing investment in developing countries, because investment is made upon returns (Pierce & Jequier 1983; Saunders 1983; Hudson 1984; Wellenius 1984; Ouko 1987; Townsend 1991). Unfortunately, many developing countries are being caught in the vicious cycle between strained investments and mounting demand for basic telephone service, so that high economic benefits can hardly materialize.

Funding shortage is a serious hinderance to China's telecommunications growth. In the 1950s and 1960s, the centralized funding was able to cover the expansion. Theoretically, as the industry grows, the demand will also increase. As demand exceeds supply, "it is easily possible with proper pricing policies to recover from the cost of providing telecommunications services, including the cost of capital and system expansion" (Saunders et al. 1983). In the case of China, this principle of supply-demand equilibrium is yet to be confirmed. In fact, it has shown just the opposite: the more the industry grows, the more the government has to invest, because the development does not generate sufficient profit margins to finance itself. This dilemma is not unusual in developing countries. It lies in the political consideration of development rather than economic priority. For example, to promote basic telephone service in a poor population, the service must be affordable and the government has to absorb high costs from providing the product and service. Raising the price to balance supply and demand in a socialist country was unheard of, because the role of government was to maintain low price regardless of market demand and technological conditions.

As a result, for 30 years, telecommunications in China was operated in deficit and relied heavily on government subsidies. Statistics released by the MPT indicate that the combined cost for telephone service using crossbar switch (semiautomatic) was 2,500 yuan ($530, then exchange rate), but the customer would pay only 200–400 yuan for installation and a flat monthly fee of 14 yuan. The cost of SPC (automatic) phone was 5,000 yuan ($1,060), but the customer was charged 1,500 yuan ($320), only about one-third of the total cost (Ji et al. 1990). It is evident that as new technologies or services are introduced, the government has to subsidize more if the financing policy should remain the same.

The central control of industry financing is a critical characteristic of telecommunications in China. It reflects the ownership, operation, and political ramifications of the telecommunications industry. Therefore, changes in financing signal a fundamental movement away from the central regime. The impacts of these changes cannot be underestimated. The next section will discuss the changes and impacts in more detail.

FUNDAMENTAL CHANGES SINCE THE 1980S

China's economic reform since the late 1970s is the most fundamental force that has brought many changes in the telecommunications industry. The significant growth began in 1979, when the government implemented the ambitious economic reform and open-door policy. The reform has produced remarkable results: the GNP increased 150 percent between 1979 and 1988 (*Business Week* 1990), and the quality of life has significantly improved in terms of disposable income, education, and entertainment consumption. The changing economic conditions have stimulated demand for telephone service both from businesses and residents and prompted reevaluation of telecommunications in its roles in the economy. At a conference on telecommunications in the Pacific, delegates from Chinese MPT provided the following conception of telecommunications infrastructure and how it should be developed:

> With the policy of reform and opening going to greater depth and the national economy keeps on developing, telecommunications as an infrastructure is assuming an ever more important role in promotion of economic development. It is instrumental in saving time and energy, speeding up financial and commodity circulation [*sic*], enhancing investment effectiveness, and relieving the reliance on other forms of communication. . . . It offers vast social benefits in general and affords great indirect economic benefits to the user. China has learned that telecommunications is an advance agent, and its

construction must take precedence to economic development. (Liang & Zhu 1988)

The words reflect a strong sense of urgency for telecommunications to better position itself in the economy. In fact, it has been made the top development priority, along with transportation and energy. This translates to many favorable policies and treatments such as taxation, construction, and distribution of foreign exchange (Sun 1990b). During the next five years (from 1991 through 1995), if China's economy can achieve the projected growth of 6 percent per year, it is reasonable to expect the industry to obtain growth of 10 percent to 12 percent in switching transmission capacities and 15 percent in telephone sets during the same period.

A high correlation between economic conditions and telecommunications development follows the above scenario. The improved and more vigorous economy requires more information that needs remote access and instant responses by telephone. For example, many coastal cities have achieved high economic growth in recent years, such as Shenzhen, Guangzhou, Xiamen (Amoy), and Shanghai. The telephone density in these regions is also the highest in the country, about 12 percent.[2] In contrast, in many rural communities where the economic activity is dominantly agrarian, the telephone density is much lower, typically below 0.5 percent, and many still use manual step-by-step switches.

The economic growth improves industry financing. The central financing has been following the proportional investment scheme based on the total annual budgets. Therefore, if the economy continues to grow at a steady pace, actual government investment should also increase at the same or a higher ratio. Moreover, telecommunications will be able to seek finances from other sources, primarily provincial governments and enterprises. As the local economic conditions are improved, local governments are enthusiastic to invest in territorial telephone systems at a much higher ratio than the central government. It has been recognized by local governments that adequate telephone systems do contribute to the economy in general.

Market Demand

The improved economy has stimulated tremendous demand for telephone services, new market dynamics that were ignored before. Despite double-digit growth in the number of switching lines and telephone sets, the demand, especially from urban residents, is soaring at an unprecedented scale. For example, there were 60,000 requests in 1986 for telephone service (Greer 1989). Three years later, the waiting list jumped to 520,000 despite a strong penetration (*World*

Telecommunications 1990). Yang Taifang, Minister of Posts and Telecommunications, acknowledged recently that the numbers have doubled in a year to "well over one million" nationwide (Li 1991). Figure 10.2 shows the growth of number of telephone sets and density in population.

The market demand comes from two sources. The first, which has increased dramatically, is urban businesses and residents. Although the urban population constitutes only 25 percent of the total population, they take up more than 80 percent of service requests. The rapid economic development in urban areas has contributed to the high demand. Despite soaring service prices, people still desperately want to have a telephone at home for communication needs. The urban demand can be further identified as local loop versus direct switched lines. Although a considerable number of calls on local loops are transported to public networks, the greatest demand is generated in local areas such as office complexes and concentrated residential areas. This type of demand explains the high sales of PBXs in the market. Moreover, demand from rural areas is also in the rise despite an annual production of 600,000 circuit crossbar switches (Sun 1990b). The rural demand comes primarily from mushrooming rural enterprises in manufacturing and exporting, as well as some wealthy rural residents. Given the size of the rural population and its accelerated activities with cities, the demand will remain strong.

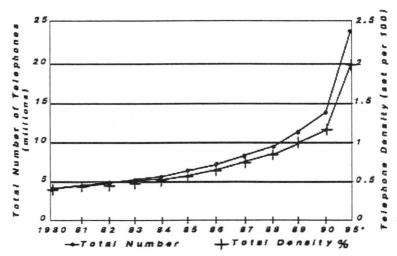

FIGURE 10.2
Growth of Telephone Quantities and Density (1980–90)

It may take well over a decade for the market demand to stabilize as telecommunications production and services gradually cover major urban and coastal rural areas. During this period, the demand for basic service will remain an annoying pressure on the government that has to be resolved. However, the demand cannot be alleviated by simply supplying more telephone sets; that will cause serious traffic congestion on switched lines, because in many cities, exchanges are already operating at near-peak switching capacity. Typically, the urban traffic congestion rate is about 50 percent during business hours. The increase in the number of telephone sets must be in accord with switching and transmission capacity, especially on public networks. The backdrop of high demand, as cautioned by Wellenius et al. (1989), is that high unmet demand diminishes the potential economic returns of telephone networks, thus affecting economic growth in the long run and the national infrastructure.

Diversified Financing

Distinct from telecommunications development in other countries, diversified financing is a unique policy in China to overcome financial bottlenecks. The massive economic reform provides a favorable environment, and conflict between rapid growth and insufficient government investment is the reason for the alternative financing policy.

For more than 30 years, China's telecommunications industry relied heavily on government subsidies, but in 1981, the MPT made profits for the first time, thanks to the economic reform (Zhang 1986). By 1989, revenue from telecommunications service alone (excluding postal services) jumped to 7.6 billion yuan ($1.6 billion), of which telephone services accounted for 83 percent (*World Telecommunications* 1990). Not surprisingly, there is a strong correlation between telecommunications revenue and industry investment (r = .91) with an advance margin of $400 million to $500 million a year (Figure 10.3). The figure also shows that during the "take-off" period (1980–85), the government increased investment significantly, but the industry revenue was rather flat during the same period. Since then the revenue has obtained consistent growth, whereas investment has gradually leveled off.

Figure 10.3 indicates strongly that if investment continues to increase the revenue should also increase. In reality, however, investment from the central government does not increase significantly. In fact, it has remained steadily at 0.8 percent to 1.1 percent of total national investment, even lower than that in some developing countries (Sun 1990a). According to MPT officials, the industry needs about 4–5 billion yuan ($1 billion) per year for basic construction, research, and service expansion, but the allotment

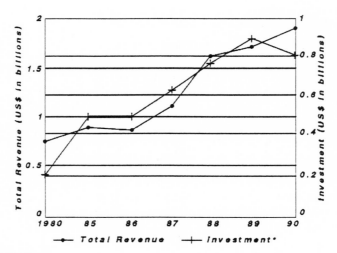

FIGURE 10.3

Telecommunications Revenue and Investment (1980–90)
*Central government sources only.
Source: Liang, X., & Zhu, Y. (1988, September). The development of telecommuni-cations in China. Paper presented at Conference on Pacific Basin Telecom-munications, Tokyo; China telecommunications services and investment in 1988. (1988a, May). *World Telecommunications, 2*, 53; China telecommunications services and investment in 1989. (1990, May). *World Telecommunications, 3*, 55; Zhang, T. (1990, February). Basic telecommunications statistics in China. *World Telecommunications, 3*, 53; Sun, L. (1991, January). A status report on China's telecommunications after 1989. Paper presented at Pacific Telecommunications Conference, Hawaii.

from the central government is only about one-tenth of the amount required (Hao 1990).

It became clear in the early 1980s that more financing had to come from other sources. With government endorsement, a new financing policy, called "diversified" or "assorted pie" scheme, has been imple-mented. Instead of one single source trying to cover a wide range of needs, the new policy allows resorting to any financial sources available, according to project, location, and responsibility. For instance, the diversified financing is now used for interprovince long-haul networks, the local governments financing the portion in their territories. Likewise, cities pay for urban telephone systems and local services. The MPT is responsible for financing only transnational networks; it no longer supports local construction, such as intercity systems (Hao 1990). Table 10.2 outlines the extent of diversified financing policy.

TABLE 10.2
Division of Diversified Financing

Types of Networks	Financing Sources
Transnational	MPT
Interprovince	Provincial Governments
	Partially MPT
Intraprovince	Provincial Governments
Intercity	City Governments

Source: Sun, L. (1990b, winter). Telecom in China: One step forward, two steps back? *TelecomAsia, 1*, 11–16.

Interestingly, central government financing has been declining in recent years. By comparison, collective funding at enterprise level has increased significantly. Table 10.3 lists the financing distribution as of 1990.

TABLE 10.3
Distribution of Telecommunications Financing

Sources	Percentage*
Central Government	32.0
Local Governments	8.0
Domestic Loans	3.5
Foreign Loans	3.0
Enterprises	50.0
Other	3.5

*Revised with updates.
Sources: Liang, X., & Zhu, Y. (1988, September). The development of telecommunications in China. Paper presented at Conference on Pacific Basin Telecommunications, Tokyo; Sun, L. (1990a, May 28). China follows the long road to telecom growth. *Telephony, 218*, 22–30.

The diversified financing has had several important impacts. First, it has reinforced local autonomy in determining projects and investment scales. Second, it has increased local investment, because the returns on investment will contribute to the local economy. Third, it has caused drastic increases in service pricing. Reports indicate that the prices for basic telephone service have increased by 30 percent to 100 percent in many cities and become an important investment source for local telephone services (Ji et al. 1990; Sun 1990b). Interestingly, the price hike does not seem to hold back the public demand, an indication of tremendous market potential and high profit margins.

Although the diversified policy has been effective, it by no means suggests an end to funding shortages. It is effective because it provides a convenient alternative to seriously strained central resources. Given the size of the China market, telecommunications financing will be a thorny problem for many years, and there does not seem to be any quick-fix solution in the near future. The problem may be intensified by market expansion and adoption of new technologies.

The pragmatic orientation of economic reform and industrial restructuring will support and improve diversified financing to sustain the growth. Specifically, the following factors will have significant impacts on the scale and speed of growth in the 1990s.

1. Local telecommunications authorities represent a large pool of provincial, city, county, and township financial sources. Increasing autonomy will encourage local motivation in telecommunications constructions and services. The increasing local investment will give localities more leverage in economic development, technical advancement, and bargaining power with the central government. Investment from local government will play a critical role in bolstering the national growth.

2. Collective financing from enterprises and communities. In the next 3–5 years, many more products will be made by local manufacturers, with or without government orders, following market demand and high profits, such as medium-capacity PBXs, terminal equipment, cables, and components. These products will be financed entirely by collective sources.

3. Foreign loans will be a vital source for acquiring advanced technologies and products despite relatively small dollar values. Foreign loans (primarily from Western governments) were the main source for purchasing turnkey systems and transmission facilities in the 1980s and made significant contributions to systems upgrading and expansion. This will likely continue. Foreign lending in some way reinforces diversified financing, because many loans are allocated for local projects, thus improving local infrastructures and, to some extent, local autonomy.

Competition

As industry financing becomes decentralized, competition intensifies. Although a generally relaxed environment contributes to both developments, competition has been more a result of industry restructuring in the mid-1980s.

In 1985, the MPT formed China Posts & Telecommunications Industry Corporation, a semi-independent holding company consisting of ten enterprise groups by industry sectors, such as microwave, telephone switching, fiberoptic transmission, and rural telephone systems (Jing 1986). Each group is largely self-contained, with manufacturers, research centers, and marketing forces. Although they are still subject to MPT's regulations, their more-or-less-independent establishment has attained considerable power in investment, technology, and management. More significantly, they have formed strong horizontal alliances among both MPT and non-MPT companies. The structural change has transformed the telecommunications industry from one supreme center to many smaller centers. The power formerly held by the MPT has been diverted to many specialized groups. As a result, the industry has become more responsive, efficient, and, not surprisingly, profitable.

The industry restructuring has fostered sector competition across the MPT, other government agencies, and companies. As mentioned before, the MPT is not the only telecommunications player in China; there are other ministries that manufacture and operate large-scale private networks. For example, the Ministry of Petroleum controls 15 satellite earth stations linking on-shore oil and gas production with voice and data transmission (Lerner 1987). Like MPT, these government entities are usually self-sufficient, with considerable know-how and capabilities. As the result of restructuring, many non-MPT companies have also joined the new industry groups, freed to compete in a somewhat open market.

The competition takes place in two forms: among MPT-affiliated companies and MPT versus non-MPT companies. The competition concentrates heavily on sectors/products that use proven technologies and are less regulated, such as SPC PBXs in medium capacity (that is, 50–200 lines), interface components, telephone sets, and coaxial cables. These sectors used to be dominated by the MPT but are now open to virtually all capable manufacturers certified by MPT's inspection apparatus (Sun 1990b). Sector competition has stimulated production for the demand. It will continue to generate high profits before the demand is largely met. Products that require sophisticated technologies still enjoy the MPT monopoly, such as CO exchanges between 500 and 2,000 circuits, fiberoptic cables, digital microwave, and satellite transmission and reception equipment. The high concentration of capital and technology in these areas has shunned competition but it may be only a question of time, as these technologies are penetrating very quickly.

The external factor for sector competition is societal demand. Because the demand still outgrows the production and service supply, it exerts pressure in the government against any significant reversal of the present policy, and that competition will likely continue its present momentum. However, the government is unlikely to take a laissez-faire approach and let it become a formidable force undermining MPT's dominance. Given China's high unmet demand for basic telephone service, MPT's continual dominance may be an effective way to control the pace of technological change, price, and equal distribution of services. If the role of MPT were totally dismantled, unbridled competition might cause inflated prices and confusing standards, which would delay the adoption of new technologies and systems development. These concerns are very sensitive and must be carefully balanced while maintaining adequate competitive scales.

The future prospect of competition is not clear, depending on how other market forces evolve, such as economic reform, demand, and technology. The government's policy seems to be to continue the relaxed policy while resisting vigorously any bold attempts that may be potential threats to MPT's position. Among others, the following may have significant influence on competition in the 1990s.

1. Financial constraints. Despite the development priority, central financing will not increase quickly from its present level. To achieve the projected growth, the industry will have to continue the diversified financing policy. This will encourage more manufacturers and suppliers to compete in the national and regional markets, with little MPT intervention.

2. Increasing horizontal alliance and local autonomy will compete aggressively because their survival is secured no longer by the government subsidy but by market performance and profits. Their gaining power will have strong impact on the market dynamics.

3. Competition will also be fostered by new technologies. Although MPT controls most technical information and R&D facilities, others may still try to gain leverage in the market by introducing new products. As development of new technologies continues to accelerate and diversify, it will create many new opportunities for competition.

Technology

Technology is an important building block in China's telecommunications infrastructure initiative. Its role as an "advance agent" cannot be achieved without effective policy and financial support. The more China is committed to economic development, the more it will rely on telecommunications technology. In fact, technology must be constantly improved and upgraded even for basic telephone services to accommodate increasing economic activities.

The technological gap between China and the developed world is alarming. In a strategic study entitled *China 2000: Status and Outlook of Telecommunications in China and the World*, inadequate technology was blamed for slow development and may hinder future economic growth (China Communications Association 1984). The study proposed new policies and initiatives from the government and industry. The R&D has been adjusted to focus on new technologies, and domestic production has increased substantially since then. In 1989, the State Council approved an MPT proposal for developing new technologies such as SPC switching and mobile communications (Song 1990). The pursuit of advanced technology will continue through the 1990s. The MPT's plan for the "Eighth Five-Year" period (1991–95) spells out the goals in digital switching, transmission capacity, network expansion, and number of telephones (Chen 1990). The indication is very clear that China wants to develop advanced technology as a vehicle to fulfill the demand for basic service while leapfrogging into advanced arenas to reduce the huge gap with other countries, the strategy pervasively adopted by many developing countries in Asia and elsewhere to gain leverage in the technology race.

The policy of pursuing advanced technology, however, is not without serious difficulties. Although new technologies can improve network efficiency and increase economic returns, they require much higher capital investment and favorable conditions. The policy may drain huge investments before economic benefits are realized. Nonetheless, technology will be the ultimate solution to improving the country's communications infrastructure and basic service. The impacts of advanced technologies on market dynamics and economic development in China will be tremendous and far-reaching.

Technological development in China is now experiencing a massive transition in three frontiers: from manual to automatic switch; from analog to digital switching and transmission; from POTS to integrated services. According to the MPT, the goals by the year 2000 are transmission capacity and traffic volume increased by sevenfold from 1980; telephone sets increased to 33.6 million, or about 3 percent nationwide (major cities about 25 percent); and open data communications networks among major economic centers and cities (Song 1990).

Digital switching/PBX or SPC switching is the fastest growing segment in technological development. Since the first SPC PBX was introduced in 1982, its proportion of total switching capacity has soared rapidly, from 0.9 percent in 1984 to 40 percent in 1989, with an annual growth rate of 60 percent to 160 percent (Table 10.4).

The digital switch market consists of two parts: imports and domestic production. For discussions on imports and related policy issues, see Sun (1990b), and these are not covered in this chapter.

The domestic production can be further divided into two categories. One is technology transfer or joint ventures with local assembly, such as Shanghai Bell Telephone Co.'s S1240 (with Belgium), and Tong Guang-Nortel Ltd. Co.'s Meridian SL-1 (with Canada). By 1990, there were about 12 companies (most of them

TABLE 10.4
Growth of SPC Switch Market

	1985	1986	1987	1988	1989
Urban switching capacity (millions)	3.37	3.80	4.65	5.56	6.68
Growth rate (%)	15.30	13.10	22.10	19.70	20.10
Urban SPC	0.12	0.27	0.90	1.47	2.80
Proportion of urban switching (%)	3.60	2.00	19.30	26.50	41.90

Source: Ministry of Post and Telecommunications (1989a). *Annual Report*. Beijing: Government of China; Ministry of Post and Telecommunications (1990). *Annual Report*. Beijing: Government of China; Sun, L. (1990b, winter). Telecom in China: One step forward, two steps back? *TelecomAsia, 1*, 11–16.

MPT affiliates) engaged in some type of joint venture and manufacturing.

The second category is Chinese made. According to MPT, by the end of 1989, there were about 100 companies manufacturing PBXs and CO switches. The typical domestic PBX products have a capacity between 50 and 200 switching lines; some make 400–2,000-line SPC switches. The annual output of domestic switches is about 150,000 circuits. With a growth rate of 15 percent, it is expected to reach 500,000 in five years.

Since 1989, the supply has been behind the demand, with considerable gaps. The estimated demand in 1990 was about 500,000 circuits, essentially digital switches. The high demand in China contrasts with the flat switch market in the West, especially in the United States, since 1987. The demand comes from local (primarily provinces and cities) needs for replacing crossbar capacity and transmission quality, more telephone sets, and emerging integrated information services. The gap between supply and demand indicates an accessible market for imports for some three to four years.

Another development aspect with high regard in China is fiber transmission networks. China's transmission media was (and still is) dominated by coaxial cables, which typically have a low capacity of 600–960 voice channels. Since the late 1980s, the MPT has taken initiatives to gradually replace coaxial cables with fiber trunks for transnational and interprovincial communications. The MPT believes that fiber networks will be the backbone of the nation's telecommunications because of its high capacity and reliability, easy maintenance, and low production costs. Fiber is also well suited for data communications, which are expected to increase rapidly in the future. In the "Eighth Five-Year Plan" (1991–95), seven national fiberoptic trunks and a dozen local networks will be deployed; by the end of 1995, the total length of the country's fiberoptic networks will reach 50,000 kilometers (31,000 miles), or about 12 percent of total telecommunications connection. The ambitious plan requires not only well coordinated planning but also high and consistent capital investment, and because this will inevitably compete with other national projects, a viable financing strategy will be critical.

So far, more than two-thirds of fiber trunks have been implemented in public switching networks (intercity and long-distance). Private networks using fiber transmission include railways, energy, and transportation control. By 1989, there were 151 fiber trunks in service, under construction, or being planned, the total length of which was about 11,940 kilometers (7,420 miles). Table 10.5 shows the growth of public fiber networks between 1988 and 1989.

Although China has made impressive effort to embrace advanced technologies, the effectiveness of policy has to be weighed with the following considerations.

1. Faced with enormous pressure in basic service, the policy on technology must be devised in such a way that combines the benefits of basic service with a grip on advanced technologies. They are both critical for the national telecommunications infrastructure. Throughout the 1990s, we will see divided efforts, primarily at MPT, to catch both ends. In terms of priority, however, the emphasis will be directed toward basic service to benefit the majority of society.

2. China has established fairly comprehensive R&D capabilities for basic and applied research. This will be instrumental to technological developments. However, funding constraints will restrain the scope of R&D projects. The dilemma that will continue to

TABLE 10.5
Public Fiber Networks

	1988	1989
Total length (km)	3,140	5,400
Urban nets	2,040	2,430
Long-distance	1,100	2,970
Equipment shares (%)		
Domestic	23	39
Imports	77	61
Cities using fiber systems		
Urban nets	36	44
Long-distance	9	16

Source: Liang, J. (1990, August). Present state and outlook of optical fiber communications in China. *World Telecommunications, 3*, 8–10.

baffle the policy makers is between advanced research, which requires intensive investment and is time consuming, and efforts for commercial products and basic services. Unfortunately, the centralized research mechanism makes diversified financing unfeasible from local governments, even if it may be located in a particular area.

3. Technology transfer will be used continuously as an effective and convenient strategy to improve current technologies and services. Essentially, technology transfer takes two forms: direct purchase of Western products and joint ventures importing manufacturing facilities. Funds for technology transfer will come from loans provided by foreign governments, development programs, and international banks. Therefore, to a certain extent, the amount of foreign lending will determine the scale of technology transfer. Foreign loans are not without preconditions in recipient countries. The Chinese government learned a hard lesson when all foreign lending was cancelled or postponed in 1989 after the government's brutal suppression of the democracy movement. The message is clear: if China wants to continue its technology transfer policy, it must maintain political stability and economic reform, although this does not guarantee that political unrest will not happen in the future.

CONCLUSIONS

Telecommunications in China will continue to surpass the overall economic growth, directed by government policies and mounting market demand. The problems faced by China's telecommunications do not necessarily suggest a deadlock; they may well have shown restructuring from a central regime to introducing new market mechanisms that will benefit industry and society in the long run. This indicates a continuous trend in decentralizing industry structure and policy making. China shares the same problems faced by many developing countries, such as fund shortages and imbalance in development; China, however, does not seem to be severely impaired. Instead, the problems have encouraged the government to devise alternative policies and practices. Horizontal alliances, diversified financing, local autonomy, and competition, among others, have become distinct characteristics of telecommunications development in China. They have been effective in achieving the telecommunications goals and should continue in the future.

A certain degree of industry decentralization does not suggest that the current central regime will resign soon, and its future prospects remain to be seen. Indeed, there has been a tendency to deregulation in manufacturing and pricing and to more entries and competition with little government interference. However, none of these has altered the industry ownership, and the government is very sensitive against any possibility of control of the nation's networks by nongovernment investors (Sun 1990b). Given China's demographics, government's ponderous influence, and the inherently weak private sector, the most viable solution may be the continual improvement of the central regime by mobilization of market forces, especially at local and enterprise levels.

Finally, a comment on the prospect of China's telecommunications development in the international context. Despite its remarkable progress, China will continue to be a marginal country in terms of investment scale, technology (research and applications), and services. As new technologies and products are accelerating in other countries, China still will be trying to provide the very basic telephone service to its huge population. The gap may become even larger as the cycle of technology becomes shorter over time. China may have to develop its telecommunications at its own pace and generate its own policy reconciling the central regime and increasing market forces. Telecommunications in China has sustained a rapid growth in the 1980s, and it will be interesting to see how the attributes discussed in this chapter affect its development in the 1990s and beyond.

NOTES

1. China has 30 provinces and municipalities, more than 400 cities, 2,200 counties, and 54,000 townships.
2. Shenzhen, a special economic zone adjacent to Hong Kong, has a telephone density of about 30 percent, the highest in the country. Beijing, the nation's capital, is about 16 percent (metropolitan area).

REFERENCES

Chen, H. (1990, June). Plan for developing posts & telecommunications. *China Market, 6*, 35.

China Communications Association. (1984). *China 2000: Status and outlook of telecommunications in China and the world.* Beijing: China 2000 Research Project.

China telecommunications services and investment in 1988. (1989a, May). *World Telecommunications, 2*, 53.

China telecommunications services and investment in 1989. (1990, May). *World Telecommunications, 3*, 55.

Greer, C. (Ed.). (1989). *China: Annual facts & figures* (Vol. 12, pp. 233–35). Beijing: Academic International Press.

Hao, W. (1990, August). Personal interviews.

Hauser, S., & Langhlin, M. (1991, January). Market entry strategies in the Western Pacific rim. Paper presented at Pacific Telecommunications Conference, Hawaii.

Hudson, H. (1984). *When telephones reach the village: The role of telecommunications in rural development.* Norwood, NJ: Ablex Publishing.

Inspection for SPC PBXs. (1989b, August). *World Telecommunications*, 2, 23–26.

Ji, Y., et al. (1990, July 31). Difficulty of operations in loss. *People's Posts & Telecom*.

Jing, X. (1986, June 17). MPT sets ten enterprise groups. *People's Daily* (overseas edition).

Lerner, N. (1987, August 24). Managing China's complex telecom industry. *Telephony*, pp. 32–39.

Li, Y. (1991, January 25). China to double number of telephones to 23.8 million in five years. *China News Digest*.

Liang, J. (1990, August). Present state and outlook of optical fiber communications in China. *World Telecommunications*, 3, 8–10.

Liang, X., & Zhu, Y. (1988, September). The development of telecommunications in China. Paper presented at Conference on Pacific Basin Telecommunications, Tokyo.

Ouko, R. (1987, September). Criteria of government decisions in developing countries concerning telecommunications investment. *Telecommunications Journal*, 5, 619–21.

The Pacific Century. (1990, December 17). *Business Week*, pp. 125–28.

Pierce, W., & Jequier, N. (1983). *Telecommunications for development*. Geneva: International Telecommunication Union.

Saunders, R. (1983, December). Telecommunications in the developing world. *Telecommunications Policy*, 7, 277–84.

Saunders, R., et al. (1983). *Telecommunications and economic development*. Baltimore: Johns Hopkins University Press.

Song, Z. (1990, November). China's communications towards modernization. *World Telecommunications*, 3, 3–6.

Sun, L. (1990a, May 28). China follows the long road to telecom growth. *Telephony*, 218, 22–30.

Sun, L. (1990b, Winter). Telecom in China: One step forward, two steps back? *TelecomAsia*, 1, 11–16.

Sun, L. (1991, January). A status report on China's telecommunications after 1989. Paper presented at Pacific Telecommunications Conference, Hawaii.

Townsend, D. (1991, January). Telecommunications infrastructure and economic development: Principles for formulating investment policies. Paper presented at Pacific Telecommunications Conference, Hawaii.

The use and development of local telephone exchanges. (1989c, November). *World Telecommunications*, 2, 10–12.

Wellenius, B. (1984, September). Telecommunications in developing countries. *Finance & Development*, 21, 33–36.

Wellenius, B., et al. (1989). *Restructuring and managing the telecommunications sector*. Washington, D.C.: World Bank.

World Bank (1990). *World Development Report 1990*. Oxford: Oxford University Press.

Zhang, P. (1986, February 5). Outlook of posts and telecom in 1986. *People's Daily* (overseas edition).

Zhang, T. (1990, February). Basic telecommunications statistics in China. *World Telecommunications*, 3, 53.

Zhou, C. (1990, February). Strengthening management and development of communications industry. *World Telecommunications*, 3, 3–4.

11

Telecommunications and Regional Interdependence in Southeast Asia

Meheroo Jussawalla

Many forces have dramatically changed the arena of international relations, but none as powerful as the telecommunications revolution. It has wrought changes in the way society functions, economies grow, and markets get integrated. Political changes and international negotiations have been transformed by dynamic telecommunications technology. Electronic highways have created an intelligent universe in which banks and financial institutions transact business in nanoseconds. Television and broadcasting are becoming interactive media that not only are creating changes in political and social systems but also are being increasingly used as tools for international diplomacy. Attempts are being made by the International Telecommunications Union (ITU) to launch the global village through telephone and satellite networks that will ensure access to telecommunications channels for even remote areas. The reason for this global demand for new technologies is the reliance that society places on information, which is now the major current in the international marketplace. Both developed and developing countries need access to information, which is both a commodity and a resource. The critical issue, however, is the control of the switches of information and their ownership. This has led to the rapid deregulation of monopoly structures that controlled these switches so that the beneficiaries of the information revolution will be the users.

In such a globally charged environment of sophisticated networks, the developing countries of the Asia Pacific region and those of Latin America are rapidly becoming participants in the new information age and are ascribing higher priorities to investment in telecommunications equipment and services than they had in the past. The wave of deregulation and privatization that has been sweeping the industrially advanced countries since 1982 has also spread to Southeast Asia, creating greater opportunities for businesses and households to avail of the new technologies at lower costs. Benefits of latecomer access have added to the flow of information, enriching human resources and infrastructures.

Investment in infrastructures to create and maintain the new information technology (IT) has been a major determinant of the growth of the macroeconomy. In order to make structural adjustments, existing investment patterns had to be altered to effect the change from capital intensive to information intensive investment. The growth of the primary information sectors of the countries of the Asia Pacific region created both backward and forward linkages with other sectoral growth and served as the leading sectors for development (Jussawalla et al. 1989). Such transformation was also aided by the operations of transnational corporations, whose subsidiaries provided the needed technology transfer to the developing countries, and their demand for telecommunications networks created further investment opportunities for the growth of new services.

The convergence of computers and telecommunications enabled the transmission of information in "real time," arousing the curiosity of television viewers around the globe and generating an unquenchable thirst for new information and new ways to link into the global networks. As a consequence of these developments, world trade has been outgrowing the gross national product of nations. International telephone traffic over the past decade has been growing at 20 percent per year. The mass media have been responsible, to a large measure, for the end of the Cold War and the desire for open markets. International relations have become far more cordial and cooperative despite the tensions in some East European countries.

Although there are both risks and opportunities involved in large-scale technology transfer, the countries of the Asia Pacific region have managed to maximize their benefits and retain their cultural identity. Japan initiated an Asian development model based on the concept of Johoka Shakai (the Information Society) that was emulated by the members of the Association of Southeast Asian Nations (ASEAN). The greatest advantage of the Information Revolution has accrued to the newly industrialized economies of Hong Kong, Singapore, Taiwan, and South Korea. The Japanese development model has influenced not only this region but also the growth of newer technologies in telecommunications, along with the institutional and ownership patterns of services. Convergent technology as an agent of change was widely proclaimed by Koji Kobayashi, the former chairman of Nippon Electronic Corporation (NEC), in his concept of C and C (computing and communications). Convergence today is best obtained through multimedia based on digital systems.

Multimedia is dependent on the combination of compact discs with superchips and fiberoptic cables. The promise of multimedia for users in offices and homes has been further strengthened by the technology of virtual reality. These changes are moving to Southeast Asia rapidly from Japan and exciting the planners of one of the fastest-growing regions of the world. Most of the ASEAN countries recorded growth rates of between 6 percent and 10 percent in 1991. A concomitant increase in investment was also witnessed because of a large diversification of user demand.

This chapter focuses on the new policies for regional integration and regional trade blocs that are emerging in Southeast Asia in response to the free trade groups like the North American Free Trade Agreement (NAFTA) and the European Single Market. The first section of this chapter examines the need for and the attention paid to this issue in public policy. The second section will analyze the restructuring of the telecommunications industry consequent to the privatization and liberalization policies undertaken in the recent past. This is followed by an overview of the standardization issue in the region and its likely impact. The last section deals with regional satellite systems and their role in the advancement of development. The conclusion gives a brief overview of alternate systems that might enhance investment in telecommunications and provide more affordable services for users.

THE REGIONALISM PARADOX

At the ASEAN meeting in 1991, Malaysia's Prime Minister Datuk Seri Mahatir Mohamad has raised the important issue of whether Third World countries that do not belong to the trade agreements of advanced countries should form themselves into regional groups for the advancement of their countries and pursue policies of free trade within such groups. The ASEAN countries have always espoused free markets and welcomed foreign investment. For them, the call of Mahatir becomes rhetoric, yet policy makers in the region are confronted with the international realities of the European Single Market and the NAFTA. Will the existence of such trade blocs impede the development of the Near NICs of the region and further add to the inequalities of the world order, or will they help promote the exports of the ASEAN countries to the regional blocs of the West? With the General Agreement on Tariff and Trade (GATT) still struggling with issues of free trade for services and telecommunications and being a part of the General Agreement on Trade in Services, it appears paradoxical that the most outward-looking economies of Asia should be considering a regional trade bloc. They have always recognized the value of free trade for unrestricted development and wealth creation, but today they also recognize the economic power of an Asian giant like Japan and its challenge to the West. Mahatir's call to "Look East" and to follow Japanese policies of export-led growth are attractive to other countries in the ASEAN.

In 1990 there was a proposal for an East Asian Economic Group, but this had not been previously approved by other members of the ASEAN. However, most countries approved the logic behind the proposal in that if the North Americans and the Europeans have the right to form economic groups and yet be members of GATT, East Asians could do the same. However, most Asian countries have larger trade balances with the United States than with Japan. For example, Malaysia's manufactured exports to the United States are 25 percent of its total trade, compared with 8 percent to Japan. The United States and Australia have been Malaysia's major trade and security partners. This is true of all the members of the ASEAN.

This proposal for a regional grouping of countries did not meet with approval at the ASEAN meeting in February 1992 and for the time being is on the back burner. The Japanese have officially distanced themselves from the concept of an East Asian Economic Caucus even though the proposal enjoys considerable support from Japanese corporations. This is also true of businesses in Singapore, Thailand, and Indonesia. Compounding this desire for regional economic cooperation is the end of the Cold War and the flow of foreign direct investment to eastern Europe. Even Japanese investment to the ASEAN countries has declined relative to five years ago as Japanese companies explore new markets in Europe. Furthermore, the concept appears superfluous in light of other regional groups that are already operating, such as Asia Pacific Economic Cooperation (APEC) and Pacific Economic Cooperation Council (PECC). The difference between these two organizations is that APEC is composed of representatives of the member country governments but the PECC is made up of business and academic representatives. There is, however, close coordination between the two groups.

In the telecommunications sector, there are already many regional collaborative arrangements that have been spurred on by multinational corporations as well as by governments. The role of governments has been highlighted by the creation of multilateral telecommunications facilities suppliers such as Intelsat, which was formed in 1964 and now has more than 120 government signatories. There are other government-sponsored regional consortia such as Eutelsat, Arabsat, and Palapa and Post, Telephone, and Telegraph (PTT)-based consortia for undersea cables that cross the Pacific and Atlantic Oceans. Inmarsat is an important international cooperative of governments for providing maritime and mobile communications.

The relationship among these institutions is not easy to define. On the one hand they compete among themselves as well as in the services they supply, and on the other hand, they compete with other sectors within their domestic economies for budget shares and priorities for investment. This gives them the characteristics of being both regional and domestic organizations.

These international and regional cooperatives have proved to be extremely profitable organizations in addition to being pioneers in the application of new and sophisticated technologies. As a result, such corporations have spurred competition by the profitable nature of their operations, as is shown by the entry of Panamsat, a regional satellite system for parts of South America and Asiasat, which has practically monopolized broadcasts to Asia.

The impact of multinational investment in Southeast Asia has also been one of creating regional interests like the consortium of KDD with Cable and Wireless and Pacific Telesis for the submarine cable that links Tokyo with Oregon. It has given rise to a regional grouping of telecommunications competitors to get a share of the growing market. Electronics-based investment by transnational corporations (TNCs) started in Southeast Asia in the 1960s with low capital assets and high levels of employment. Such investments served as a catalyst for developing the service sector. Both Hong Kong and Singapore developed their financial and banking services and established electronics facilities for their stock exchange operations, going into global round-the-clock trading. These "intelligent cities" led the way for regional flow of services in the banking and financial sectors that further helped the development of financial services in Taiwan and South Korea. Preferential tax privileges also promoted to the flow of foreign direct investment to East Asia, so that the total assets of these conglomerates located in the region were 10 to 15 times the GDP of their host countries. Much of this growth is attributable to the supply of telecommunications services that enabled speedy electronic funds transfers. In turn, this led to faster systems of data transfer, establishing interdependence among the leading exporters of the region.

Joint ventures between U.S. and Japanese corporations in the telecommunications sector have played a significant role in restructuring corporate assets. These companies find such mergers to be a low-risk method of entering global markets. For example, American Telephone and Telegraph (AT&T) struck a deal with NEC in 1990 to trade its computer-aided designs for NEC's advanced logic chips. Likewise, Texas Instruments combined with Kobe Steel to manufacture logic chips in Japan. Motorola has joint ventures with Toshiba and Hitachi, INTEL with NMBS, and Micro Systems with Sony. This spate of joint ventures shows how regional trade is already existing without formal agreements. Corporate joint ventures show how the transnationals are bypassing government negotiations and lend support to the hypothesis of Kenichi Ohmae in *The Borderless World* (1990) that corporations bear greater loyalty to their customers in different parts of the world than to their home governments. Even China, which does not permit TNCs to locate their subsidiaries within its territories, still takes advantage of foreign participation through joint ventures. Alcatel of France, Siemens of Germany, and NEC of Japan are all providing stored program control exchanges under such agreements.

PRIVATIZATION POLICIES AND RESTRUCTURING THE INDUSTRY

The United States first initiated the policy for opening the markets to competition for its monopoly suppliers through the divestiture of AT&T in 1982. In 1985 Japan followed with a new law to deregulate the Nippon Telephone and Telegraph Company (NTT) and was preceded by similar changes in the United Kingdom. The major reason for these policy changes was the push from new technology and the enormous demand from users to avail of the new technology. Regulators the world over found that they were unable to keep abreast of these dynamic changes in telecommunications and their convergence with computers that rendered their regulations outmoded. In Southeast Asia in general, there was the practice of having statutory bodies called PTTs. These controlled all the services in telecommunications, and their profits were diverted to the general government budget, so their investments got shortchanged. Because of the wave of deregulation in Western countries, these PTTs realized that the time had come for them to modify their regulatory practices and permit some opening of the markets to competition, especially in value-added services.

The Asia Pacific countries do not strictly adhere to either the U.S. or the Japanese model for liberalization of their PTTs because of diverse political doctrines and economic organization. Within the region, policies for deregulation vary from a fully centralized system, as in Singapore, to a mix of public and private sector operations, as in Malaysia. The need for technology transfer also brings pressure for change, but the ultimate decision depends on domestic factors and consumer demand. Policy makers in these countries are becoming increasingly aware that deregulation by itself does not obtain network efficiency; it has to be accompanied by organizational and structural changes and a policy for transition management. The implications are twofold and interrelated. First, restructuring of telecommunications impacts on all other sectors of the economy that are dependent on its services. Second, liberalization policies have to be compatible with global market trends to enable these countries to avail of new market opportunities in other parts of the world. The ASEAN region realized that neoclassical interventionist policies followed by other Asian countries impeded economic growth, and they chose to follow free market policies instead.

The estimated market size for telecommunications equipment and services in East and Southeast Asia combined was about $113 billion for 1990 and is expected to increase by 10 percent per year for five years to $178 billion by 1995. This has resulted in growing competition from multinational vendors who are trying to gain a foothold in these markets. Infrastructure for both basic and enhanced services is being competed for in the region. In 1991, Indonesia called for bids to expand 350,000 new telephone lines, and both AT&T and NEC were competing for these bids. Political pressures were brought into the decision. As a result, Indonesia doubled its plan and split the bid between the two. At the same time, British Telecom won a bid to install 300,000 new lines in Thailand. Indonesia is also permitting private companies to provide very small aperture terminals under the country's Sixth Five Year Plan, called Repelita VI, which also will provide more telephones to rural areas to meet the goals of the Maitland Commission's report (1984), under which there should be a telephone within an hour's walk of every village. Perumtel will increase automatic telephone lines from 910,000 to 2.3 million by 1994 and to 7 million by 1997. A state-owned monopoly called Industri Telekomunikasi has a joint venture with Siemens to manufacture digital public exchange switches and pay phones with Bell Manufacturing of Belgium.

Malaysia has also privatized its PTT, called Telekom Malaysia. Its shares were placed on the stock exchange in Kuala Lumpur in 1991. This deregulation provided a window of opportunity for new service suppliers from the private sector, such as Celcom and Atur 800, both competing for the cellular telephone market. Likewise, data services are provided by two firms, Teledata and Maypac. The privatized corporation Telekom Malaysia also provides cellular telephones and telepoint, but the largest share of the market for cellular telephones (53 percent) has been captured by Celcom.

The newly formed private company Telekom Malaysia provides basic domestic and international services and has planned to spend M$12 billion on new equipment in the next five years, for which it is contacting foreign suppliers. It has called for bids to provide M$2 billion worth of new digital telephone lines. Such large investments have led to the formation of indigenous companies in the telecommunications sector, like Sapura and Federal Cables, to compete for them. Sapura already provides terminal equipment to the users.

In Thailand there are two state-owned monopolies — Telephone Organization of Thailand and Communications Authority of Thailand. The former controls domestic communications and the latter international communications. Privatization is not foreseen in the near future. Only liberalization of new services is being attempted, with the permission of the government. In July 1992 the government of Thailand announced that it will contract for an additional 1 million telephones for rural areas. A private consortium was assembled in the country, called Thai Telephone and Telecommunications, and is likely to win the contract. This consortium is led by Loxley (Bangkok), a major private operator in Thailand for satellite services. Both in Indonesia and in Thailand, basic services have to be supplied under state subsidy because of the low per capita incomes prevailing in the region. The necessary income elasticity of demand is not present in the rural areas of both these countries to support commercially provided basic services. It becomes necessary for cellular telephones to raise the low penetration ratios of telephones in the rural areas. Hitachi launched a cellular 900 phone network in Thailand in September 1991, but it caters mainly to Bangkok. Another supplier of cellular phones is Amps 800, which finds demand proliferating too rapidly. The 800 and 900 numbers following the names of the suppliers implies the megahertz frequency bands that will be used for the mobile services.

Singapore has been described as the "intelligent city" of Southeast Asia because it has provided the latest and most sophisticated information technology to its citizens. These services were provided by Singapore Telecoms, which has been a state-owned organization with complete monopoly over every new service supplied. Both domestic and international service along with paging, cellular, and videotext have been under the same monopoly. Also, it has a statutory board called the National Computer Board for the computerization of the country, which provides a computerized network to support the most important activity of the republic, namely, trade. The Tradenet system documents all trading activities of its busy harbor. The government continues to direct media policy even though it has permitted CNN and ESPN to provide cable television programs. The Singapore Press Holdings, which held the monopoly over newspapers for decades, will now face a competitor (albeit from the government sector), that is, the National Trade Union Congress, which will publish a rival paper to the Straits Times in 1993. Singapore has become a major investor for information technology equipment in Sri Lanka, Saudi Arabia, and, to some extent, the Silicon Valley. The investment in the United States is not as profitable as anticipated. Now Singapore finds that it has to bow to the winds of change and has declared that it will privatize its telecommunications services in 1993 and place the share of Singapore Telecoms on the stock exchange.

Regulatory changes in South Korea have been slow in coming. Monopolies of the state-owned telecommunications operator Korea Telecom are being opened to competition. In 1991 the Data Communications Corporation of Korea started to provide international telephone services and data services in competition with the Korean Telecommunications Authority. The government also planned to license a new operator for cellular communications, but the presidential elections in December 1992 led to the postponement of the mobile communications contract. A consortium of three foreign firms — GTE of the United States, Vodaphone of the United Kingdom, and Hutchison of Hong Kong — had joined with Sunkyong Corporation to bid for the system. In semiconductor chips, Korea has caught up with exports from Japan to the U.S. market, but it has a private sector that is more oligopolistic than in Taiwan and Hong Kong. Both Samsung and Lucky Gold Star have carved out niche markets for supplying electronic equipment to ASEAN and U.S. markets.

Taiwan has two regulatory authorities, namely, the Ministry of Communications and the Directorate General of Telecommunications. Since 1987, Taiwan has progressively lifted regulations to suit its political imperatives. Competition was introduced in customer premises equipment and domestic value-added networks. Taiwan has also taken the lead in imaging technology and in the manufacture of mobile telephones and personal computers.

Perhaps the most important deregulated private sector enterprise in Hong Kong has been the Asiasat system. Hong Kong's telecommunications services are modern and enable the country to serve as the regional center for global business and to be a gateway to China. Almost 80 percent of its central office switches are digital. Its competitive and aggressive pricing of services has given the country an excellent telecommunications infrastructure. Hong Kong Telecom provides basic domestic and international service, and its major shareholders are Cable and Wireless and China International Trust and Investment Company. It has a large mobile communications sector that grew by 13 percent between 1991 and 1992. There is competition in the cellular market from Hutchison Telecom and Pacific Link. Hutchison has been the leading provider of cellular services since 1985, with a 50 percent market share. It is anticipated that this market will grow exponentially after 1995, when the remote areas and the new urban centers of China will demand cellular services, which will be more cost effective than terrestrial services over large land masses. It has been suggested that if resale of international circuits could be legalized in Hong Kong, there will be grater competition and more rational pricing of international telecommunications services in Southeast Asia (Muller 1992).

Adding to the consumers' choice in telecommunications services in the region is the extension of the TransPacific fiberoptic submarine cables linking continents and reducing tariffs for international services. Since mid-1990, there has been in operation a submarine cable connecting Japan, Hong Kong, and South Korea, installed by Cable and Wireless. The North Pacific Cable links Tokyo with Pacific City in Oregon and leads on to Anchorage. The Trans Pacific Cable 3 joins Hawaii with Japan and branches to Guam, the Philippines, and Taiwan. Approximately 15 cable systems are being laid in the Asia Pacific region at a cost of $3.5 billion.

STANDARDIZATION FOR A GLOBAL MARKET

Divestiture is difficult because equipment standards compatibility is essential for the efficiency of telecommunications networks, but it has not been realized, resulting in regional standards formation and competition between proprietary and negotiated standards. The term "standards" implies a uniform set of agreements between telecommunications systems and between computers made by different manufacturers. Policy makers want to protect the rights of users and make technology more affordable for them, which requires enforcing uniform standards and interconnectivity of equipment. Europe adopted the Open Network Provision in June 1990 to promote competition, but the transition to it will need agreement on standards. The ITU set up a Study Group XVIII to make recommendations for digital networks and integrated services digital networks (ISDN). The process was entrusted to the Consultative Committee for International Telephone and Telegraph (CCITT), a subcommittee of the ITU. At this point, regional standards entered the global scene, which in the past was dominated by the CCITT.

So far, nations have been collaborating in regional groups to examine this strategic issue. Europe first established the European Telecommunications Standards Institute to research uniform standards for the European Community. The GATT, in its Uruguay Round, has been dealing with standards as essential for the charter on services so that equipment and services can be traded freely. Political influences have entered the debate because of tariffs and principles governing leased lines for networks.

For the Asia Pacific region, the main issue is how to make the transition from technology-driven demand for standards to a market-driven one based on user needs. North America established the T1 Subcommittee on standards, made up of business users and vendors, and Japan has two groups for this purpose, the Telecommunications Technology Committee, made up of private corporations, and the Telecommunications Technology Council, a government body attached to the Ministry of Posts and Telecommunications. The fate of the developing countries in Southeast Asia becomes precarious because they have to accept the equipment of competing vendors, which may not be interoperable and would affect their efficiency. The very concept of open access to telecommunications systems is challenged by the competition in regional standards.

The United States held a conference in Fredericksburg in February 1990, dubbed the Standardization Summit, to establish a partnership with the CCITT to introduce a global standard for interconnectivity. The ASEAN countries are watching these developments with keen interest because of the proliferation of intelligent systems like Automatic Call Distribution, Automated Technical Complaint Service, and Credit Card Authorization using intelligent nodes to the network. With the growth of cellular mobile telephones in the region, base stations become necessary. Such mobile systems are of significance because personal communication networks reduce the cost of transmitting information at 566 megabits per second. The main impact on the Asian countries arises from vendor recognition of protocols like X.25 and X.400. With limited resources of hard currency, it becomes difficult for these countries to purchase equipment from the least costly vendors from different countries, because their products may not be compatible and may, therefore, reduce the efficiency of the entire network. Japan has exercised a major influence on the ASEAN region in this field by establishing an Asian ISDN Council in regard to the establishment of standards. Many Asian countries, including India and China, have joined the Council. In 1989, Thailand initiated a packet switched data communication network and a dedicated channel to handle facsimile exactly comparable to Euronet and Internet. Currently, it operates only within neighboring countries, but it may be extended internationally, depending on compatibility of equipment. Currently in the Asia Pacific region, two protocols are being used for electronic data interchange (EDI): one, used by Singapore, is called UN/EDIFACT and is also used by Hong Kong, Australia, and New Zealand, but Korea uses ANSI X.12 (American National Standards Institute). Likewise, for videotext, Singapore uses Teletel, Japan uses NAPLPS, and the United States uses ASCII in general. Therefore, accessing data bases across continents becomes costly and difficult. Conversion costs of equipment, including video cassettes, are high.

In order to smooth out the kinks in this problem the APEC has a special subcommittee on telecommunications that is focusing on the issue of EDI and standards. The exports of the Asia Pacific region to Europe and North America become jeopardized for lack of global standards. Consequently, the lead in standard setting for the region is being taken by Australia, and the next APEC meeting will be held in Sydney specifically to study this issue. With broad band Integrated Services Digital Network within reach of the users, the interface between computers and communications lines is driving a restructuring move on how people and machines communicate with each other.

THE SATELLITE WARS IN THE PACIFIC

Indonesia was the first Third World country to own and operate its domestic satellite and to recognize the cost insensitivity to distance of satellite communications. The Palapa system was first launched in 1976 to provide television and telephony to the 13,000 islands of the archipelago. The Palapa B2R system provides transponder capacity to Thailand, Malaysia, Singapore, the Philippines, and Papua New Guinea. In 1991, Indosat and Perumtel, the two statutory bodies governing telecommunications in Indonesia, set up a private corporation, called Palapa Pacifik Nusantra, using a new allocation for its A2 system with coverage of the Pacific Islands and Hawaii. Both CNN and ESPN have opted to use this new system for their broadcasts to Asia, in competition with Asiasat. Home Box Office and TV New Zealand are also negotiating for transponders on the same system. The charge for leasing a transponder on the Palapa system already has risen from $800,000 to $1.1 million per year. On May 14, 1992, the Delta rocket placed Indonesia's B-4 satellite in orbit for the country's state-owned company, now called PT Telekomunikansi Indonesia. It joins three Palapa B satellites already in orbit but has a longer life span of nine years.

At present, ownership of satellite receivers in Malaysia, Singapore and Thailand is illegal, and other countries in the region require licenses. For northern Asia, Intelsat provides CNN programming, but the difference is that an Intelsat satellite dish has to be at least 7.5 meters, whereas for Palapa, small dishes can receive the signals, even if they are 2.5 meters wide.

Malaysia and South Korea have both placed orders with Hughes for construction of their domestic satellites despite the availability of approximately 600 transponders by the year 1995. Not since the launch of Intelsat's first satellite, called Lanibird, in 1965 have so many satellite wars erupted in the Pacific. Aussat was launched in the early 1980s, followed by JCSat of Japan, and all these coexisted with Intelsat to provide point to multipoint communication in the region. The reason for the new competition is that traffic levels today are overtaking those over the Atlantic Ocean region and are increasing despite the growing number of submarine fiberoptic cables. The primary services are switched telephone services and long-term television leases. Intelsat will deploy its series seven spacecraft, and two of these will be in orbit by 1993. There is a growing demand for Intelsat's business service, and the Pacific Islands use its Vista service for data transmission.

With so much transponder capacity available, other companies, such as Panamsat and Pacstar, also plan to place new satellites over the Pacific in the mid-1990s. Even if demand grows exponentially in the region, there will still be surplus transponder capacity in this region, and it is not known why Malaysia and South Korea have contracted with Hughes for domestic satellites for their countries. Perhaps political considerations override economic ones.

CONCLUSION

We have observed throughout this analysis that telecommunications have become a major priority for investment and a significant contributor to economic growth in Southeast Asia. It has also become apparent that these countries have become far more interdependent for trade, investment, and technology transfer. The growth of regional trade blocs appears to counter the spirit of the GATT negotiations, and Asian exporters are now apprehensive about the NAFTA area, because Mexico will now be a major supplier of low-wage exports to the United States and also a bigger recipient of U.S. investment. Currently, Asia exports $200 billion worth of goods to the United States per year, and political pressures are mounting for a similar free trade bloc among Asian countries. They are concerned over the "rules of origin," under which goods will qualify for duty-free admittance to the North American market, and this applies to telecommunications equipment, computers, components, and electronics exports. It is anticipated that Mexico's exports to the United States could overtake those of Japan and affect the flow of data and international value-added networks as well as technology transfer operations for Asian countries. Although it is true that NAFTA brings into existence a $6 trillion market, entry to that market for most countries of the ASEAN region will become more restricted because of local content laws. However, there are countries that are planning investments in Mexico. Hitachi has moved the manufacture of video cassette recorders to Mexico from California. Likewise, Hong Kong plans to enter the Mexican manufacturing sector for its textile and electronics production. Labor costs in some countries like China and Indonesia are still lower than in Mexico. The fact remains that regionalism requires a great deal of structural adjustment, and telecommunications technology will render such changes easier to achieve. The collaboration that now exists among the ASEAN countries as well as the APEC members is achieving, more or less, the purposes of a free trade area, and regional and national satellite systems have brought a greater sense of interdependence among these countries. They are not likely to lose their lead among developing countries for participating in the information revolution.

REFERENCES

Jussawalla, M., Lamberton, D., & Karunatue, N. (1989). *The cost of thinking: Primary information sectors of ten Asia-Pacific countries.* Norwood: Ablex.

Maitland, D. (1984). *The missing link.* Geneva: International Telecommunication Union.

Muller, M. (1992, June). Telecommunications in Hong Kong after 1977. Paper presented to the ITS Conference in Sophia Antipolis, France. Mimeo.

Ohmae, K. (1990). *The borderless world.* New York: Harper Business.

12

Maximizing Benefits from New Telecommunications Technologies: Policy Challenges for Developing Countries

Heather E. Hudson

TELECOMMUNICATIONS IN THE DEVELOPMENT PROCESS

Telecommunications is a "missing link" in much of the developing world, as the Maitland Commission noted (International Commission, 1985).[1] The telecommunications link is not simply a connection between people, but a link in the chain of the development process itself.

There is now considerable evidence that telecommunications contributes to socioeconomic development. Several studies sponsored by agencies including the International Telecommunication Union (ITU) and the World Bank have shown that telecommunications can facilitate many development activities, including agriculture, industry, shipping, education, health, and social services (Saunders et al. 1983; Hudson 1984; ITU 1986). These studies have been augmented by recent research on rural telecommunications in the United States (Parker & Hudson 1992).

Distance represents time, in an increasingly time-conscious world. In economies that depend heavily upon agriculture or the extraction of resources (lumber and minerals), distance from urban markets has traditionally been alleviated only by the installation of improved transportation facilities, typically roads. Yet, transportation links leave industries without the access to information that is becoming increasingly important for production and marketing of their commodities.

Another disadvantage faced by many developing countries is economic specialization. As they strive to diversify their economies, timely access to information becomes even more critical. For example, in a manufacturing environment that increasingly depends upon out-sourcing of component parts and just-in-time delivery to assembly plants, the farther the production facilities are located from markets, the more some other economic characteristic (for example, lower wages) must compensate for the time/distance penalty.

As developing countries also join the global market by attracting multinational corporations, establishing joint ventures, and developing service industries, they soon recognize the need for a reliable and modern telecommunications network.

Telecommunications is also vital to the emerging information sectors in developing regions. The great distances between the major research institutes and development centers and the vagaries of postal services and expense of airfares mean that experts are isolated from each other as well as from the people they are trying to help. For example, the National Research Council points out that for Africa, sharing information is vital if Africans are to contribute to finding solutions to their own development problems:

Economic development in Africa will depend heavily on the development of the information sector. Countries will need the ability to communicate efficiently with local and overseas markets to determine where they may have comparative advantages for supplying their products to consumers or to purchase essential imports, based on current prices and services. Many of the economic development problems facing African countries have scientific and technological components that will require solutions to be developed in Africa by African scientists. . . . Lack of information is a critical constraint. (National Research Council 1990)

The newly industrialized countries in Latin America and Asia have commercial information sector activities that are even more dependent on fast and reliable transmission of information. For industries specializing in the provision of information goods and services, a reliable telecommunications infrastructure can make location and distance irrelevant. Singapore, for example, has become a major financial and trading center. Taiwan and Korea are now major suppliers of electronic equipment to world markets. Brazil has specialized expertise in satellite communications and aviation technology. Some developing countries have attracted "back office" information industries. American Airlines' ticket data is key-punched in Barbados; legal data for the Lexis database is entered in South Korea; Indian engineers transmit software code to Texas Instruments from Bangalore. Entrepreneurs in these countries need instant access to global information to monitor market trends and to keep up with the most recent technological innovations.

To summarize, the ability to communicate instantaneously can facilitate the development process by increasing efficiency (that is, the ratio of output to cost); effectiveness (the extent to which development goals are achieved); and equity (that is, the distribution of development benefits throughout the society).

NEW TECHNOLOGIES AND SERVICES

Fortunately, recent innovations in telecommunications and other information technologies have resulted in new equipment and services that are particularly suitable for developing country applications. The following are examples of these technologies and services.

Thin Route Satellite Earth Stations

The advent of small, low cost earth stations such as those used for rural telephony with domestic satellites and the VISTA terminals used with INTELSAT satellites bring voice and data communications to isolated regions such as deserts, jungles, mountainous regions, and offshore islands. These earth stations may be installed in any community or project site without the need for expensive terrestrial links to the national network. They may serve the surrounding territory through line-of-sight radio links (Hudson 1990).

Data Broadcasting

The flow of information within the developing world has been hampered by the cost of distribution and by the lack of access to telecommunications facilities in rural areas. Very small aperture terminals (VSATs) now make it possible for wire service information to be disseminated to virtually any location. Wire service copy is transmitted by satellite from a hub earth station, which may be shared by other data, voice, and video customers. These "micro earth stations" may be powered by photovoltaics or portable generators (Parker 1987).

Reuters uses this VSAT technology for news service feeds to Latin America. In Asia, the World Broadcast Service (WBS) based in Hong Kong uplinks news service feeds to INTELSAT's Indian Ocean satellite, which covers 80 percent of the world's population. The first customer for the WBS in China's Xinhua News Agency.

VSATs for Interactive Data Communications

Microcomputers or terminals linked to mainframes via interactive VSAT technology can be used to collect and update information from the field. A VSAT network called NICNET, operated by the Indian government's National Informatics Centre, now links 160 locations and will be expanded in the next stage to more than 500 (Blair 1988). Similar systems may be used for electronic banking — whether linking teller machines to computers or remote bank branches to headquarters — and for other interactive applications such as reservation systems, weather and pipeline monitoring, and other field data collection.

New Radio Technologies

Advances in radio technology such as cellular radio and rural radio subscriber systems offer affordable means of reaching less isolated rural customers. These technologies make it possible to serve rural communities without laying cable or stringing copper wire (Parker et al. 1989).

Facsimile

Another technology with widespread development applications is the facsimile machine, which enables any type of hard copy, including print, graphics, and handwritten messages, to be transmitted over a telephone line. Also, "fax boards" may now be installed in personal computers to allow a message created on a personal computer to be sent directly to a facsimile machine.

Audio Conferencing

A thin route service with considerable promise for development applications is audio teleconferencing. Several sites can be linked together through a bridge at a switching point or through assignment of a common frequency on a satellite audio channel.

Electronic Mail

Communication via computer is a means of exchanging information immediately. Microcomputer users worldwide may now interact using various electronic mail networks. Messages may be sent from one computer to another by communication through a host computer that is equipped with communications and message processing software including "mail boxes" for subscribers. These services are cheaper than voice communications and overcome the time zone differences that hinder real time communications. Users may dial into local nodes of packet-switched networks to reduce transmission costs. Specialized electronic mail networks have been established for developing country users (International Development Research Centre 1989).

Computer Conferencing

Another application of computer communications is computer conferencing, that is, interaction of many users through a central host computer. Each conference member may share ideas with the others and respond to their comments. Participants may log on at their convenience, thus avoiding the need for scheduling to accommodate individual schedules and time zone differences.

CD ROM (Compact Disc, Read-Only Memory)

Information in the form of data bases, full text of journals, video images, and other graphics may now be stored on compact discs and retrieved with a relatively inexpensive reader attached to a microcomputer. CD ROM's advantages include vast storage potential, low cost, durability, and ease of use. In addition, CD ROM can be used on a standalone basis, without the need for online access to data bases. Of course, the discs must be updated frequently to keep information current.

Access to Online Data Bases

Although data bases in CD ROM format are proliferating rapidly, much specialized information may be more readily available through access to online data bases. A user with a computer terminal and modem can dial into a data base and conduct a search for information specified using key words or phrases. Relevant information is then displayed in the form of citations, abstracts, or, sometimes, complete text of documents. The user can then select the relevant information and download it into the computer or print it out.

Costs may further be reduced if these searches can be localized. Data bases may be downloaded onto computers within the country on tape, floppy disks, or laser discs, with updates transmitted at regular intervals using telecommunications. The search then becomes local, without the cost of connect time.

Desktop Publishing

Enhanced graphics capabilities of microcomputers now make it possible to produce newsletters and other printed materials without typesetting. These features may be particularly valuable in countries where newspapers, texts, and development materials in local languages may be scarce and costly to produce. Development agencies may now produce their own materials in-house. Storefront desktop publishers may enable many small users to share the desktop publishing equipment and software. Desktop publishing may be combined with telecommunications (for example, facsimile), so that publications may be inexpensively produced and distributed.

APPLICATIONS FOR DEVELOPING COUNTRIES

These new technologies offer many opportunities to overcome the barriers of distance that make information acquisition and dissemination so difficult in many developing regions. The following are some examples.

Electronic Messaging

Facsimile transmission and electronic mail may be particularly viable alternatives to sending hard copies of correspondence and documents through the mail where service may be slow or unreliable. Managers and researchers located in different cities may exchange information, but these technologies can also be used to link project staff in the field with each other and with headquarters.

Electronic Meetings

Managers, development experts, or project staff may now stay in touch electronically rather than having to travel for face-to-face meetings. Audio conferencing allows participants at several sites to participate in the same meeting, and computer conferencing allows for interaction among group members at their convenience by reading and contributing to a discussion stored on a host computer. These electronic meetings do not offer the richness of face-to-face interaction, but they may be particularly important to supplement travel to meetings when transportation costs severely strain limited travel budgets.

Training

Audio conferencing may be used to update field staff without bringing them to the cities for training. For example, in Peru, the Rural Communication Services Project linked seven rural communities, three via satellite and four via VHF radio and then via satellite to the national network. More than 650 audio teleconferences concerning agriculture, education, and health were carried out during the project (Mayo et al. 1987).

Distance Education

Audio conferencing may be used to reach isolated students who may be studying by correspondence. For example the University of the South Pacific uses a satellite-based audio conferencing network to provide tutorials to correspondence students scattered in ten island nations of the South Pacific. The University of the West Indies also offers instruction to students at extension centers throughout the Caribbean using a combination of satellite and terrestrial audio links (Hudson 1990).

Access to Data Bases

Computer terminals or personal computers with modems linked to the telecommunications network can provide access to data bases anywhere in the world. Agricultural researchers, for example, may access the Food and Agriculture Organization data bases in Rome. Health researchers may search the data base of the National Library of Medicine in Bethesda, Maryland. Others may search specialized development data bases, such as those for agriculture and energy in India and for development project management in Malaysia.

Electronic Transactions

Computers combined with telecommunications enable organizations to conduct business from virtually any location. Banks may transfer funds internationally using the SWIFT network (Hudson & York 1988). Airlines may book reservations from ticket offices, airports, and travel agencies. Brokers and traders may buy and sell coffee, soybeans, copper, petroleum, and so on electronically. With reliable telecommunications links, these activities need not be limited to cities. Agricultural cooperatives may use computer terminals to find where to get the best prices for their crops. Tourist lodges in scenic areas may book reservations.

Dissemination of Information

Information for use in publications may be transmitted from the field and from regional centers to desktop publishing locations via telecommunications networks. For example, development workers and reporters in the field could send in reports by facsimile; these materials would then be edited and published in newsletters in the city. Posters and notices could be faxed to the rural communities. Newsletters could be faxed either directly to the communities or to regional centers for duplication and dispatch to schools, clinics, or government offices in their territory. Information obtained from various sources such as news services, data bases, and teleconferences could be disseminated to development workers throughout the country or region via facsimile.

IMPLICATIONS FOR TELECOMMUNICATIONS PLANNING

New and Changing Demands for Services

This proliferation of technologies and services has several implications for telecommunications planning. First, there are likely to be new and changing demands for telecommunications services.

Voice and Data

Although basic voice communication is still the first priority, many users now have requirements for data communications as well, particularly facsimile and relatively low speed data communications. Thus, transmission channels must be reliable enough to handle data as well as voice traffic.

Urban and Rural

The availability of relatively low cost radio and satellite technologies for serving rural areas makes it possible to reach even the most remote locations and to base priorities for service on need rather than proximity to the terrestrial network. However, these links also need to be highly reliable if rural users are to take advantage of the applications of facsimile and data communications described above.

Responding to Demand for New Services

The new technologies now available are likely to create demand for new services such as packet data networks and audio conferencing networks. Telecommunications managers must be able to respond to these demands both by providing the technical facilities and by setting realistic tariffs if users are to take advantage of the information sharing potential now possible via telecommunications.

Integrating Planning across Sectors

For these new technologies to serve development goals, communication planning must be integrated with national planning. If the country intends to open up new areas for settlement or resource development, telecommunications facilities will be required. If the country intends to diversify its economy, it will need to ensure that adequate infrastructure is in place. It may also need to upgrade the skills of its work force, perhaps using instructional technologies.

Lack of coordination between communications and other sectors results in wasted resources and lost opportunities. Some countries have been unable to attract new industries because they lack the necessary telecommunications infrastructure. Too often, telecommunications planning is done in isolation without information about government development priorities or new economic activities.

New technologies allow telecommunications planners to respond to changing needs by installing radio or satellite links, for example, to serve new customers or development projects, yet planners often do not take advantage of this new technical flexibility by authorizing modifications to existing network plans. They may also hinder development even if the equipment is in place if they enforce unrealistic technical standards or adopt tariffs that make it virtually impossible for potential customers to take advantage of the newly installed facilities.

User Involvement

Users are rarely heard from when the telecommunications plans for developing countries are being prepared. Yet, the users are the most important element of any plan; without an understanding of their needs and constraints, telecommunications services may be inappropriately designed or priced. Why are users so often silent? They may not have the technical expertise usually expected in planning activities; they also may be unaware of how and when to get involved.

In a sense, telecommunications planners have to act like extension agents to get out and meet with users, learn about their needs, and help them translate their requirements into facilities and services. This is a new role for telecommunications carriers worldwide. It is a particularly important function in developing countries, where resources for new facilities are limited and failure to meet user needs can hinder the economy as well as limit the carrier's projected revenues.

To summarize, for coordinated communications planning to occur:

Telecommunications administrations must be informed about national priorities and development plans.

National planners must be made aware of the importance of telecommunications infrastructure to national development.

Resources for extension and improvement of facilities must be allocated to the communications sector, and resources for training and utilization of facilities must be included in the sector budgets.

Potential users must be made aware of the services available and how they could benefit from them.

BOTTLENECKS AND BYPASS

What if telecommunications administrations are not responsive to customers or innovative in offering new services? The following sections examine how telecommunications administrations often act as bottlenecks that impede access to facilities and services and what happens when users get frustrated with trying to go through the bottleneck and decide to go around it instead.

The Post Telephone and Telegraph Administration Bottleneck

Today about 80 percent of the world's population has no access to reliable telecommunications, despite the availability of VSATs and other relatively low cost technologies (VHF, UHF, rural subscriber microwave systems, cellular radio). Given the demand, the reduction in costs, and the potential benefits, why is the diffusion rate so low? There are many reasons, but the major problem is that in most cases the Post, Telephone and Telegraph Administration (PTT) acts as a gatekeeper or bottleneck that prevents customers from obtaining equipment and services. Thus, the government-operated utility model that was adopted to protect the public interest now acts as a constraint to retard growth of the telecommunications sector and, as a result, the economy as a whole.

Two examples illustrate the result of the bottleneck. First, the number of television sets has greatly exceeded the number of telephones in most developing countries. There are as many as four to five times as many television sets as telephones in some developing countries (Parker et al. 1989). This imbalance is generally not the result of government priorities but of lack of awareness of the role of telecommunications, lack of coordinated planning in delivery of services, or unwillingness to give up political and economic control.

In a second example, we can compare access to microcomputers with access to VSATs. The reduction in cost and increase in computing power of microcomputers have been accompanied by advances in telecommunications that should make affordable voice and data communications available virtually anywhere. However, access to computers is through a competitive marketplace, whereas telecommunications services are provided in most countries through access to a single national network.

In order to understand the diffusion of these technologies, it is necessary to identify the various participants and their roles in the diffusion process. A microcomputer is a standalone technology (despite the fact that it can be linked to other computers through a network), but a telephone set and a VSAT are useless by themselves — they require a network. Thus, users cannot meet the need for telecommunications just by buying a telephone set, a cellular phone, or a VSAT. They must be able to connect to a network, and the PTT controls access to the only network. If the PTT is not responsive to consumer needs, as is often the case, frustrated consumers remain without service. These same consumers can usually buy standalone technologies such as computers and videocassette recorders (VCRs) on the open market.

Bypass: The Users' Response

When users are unable to obtain the capacity they need or afford to use available services, they look for alternative solutions. In the old days (and still today in some parts of the developing world), they turn to high frequency (HF) radio. HF is frustrating in its signal quality and varying reliability, but the price is right. If the users own their radios, they can use them whenever they want without paying a carrier. Now, satellites offer a more reliable bypass option.

Two examples of public service satellite bypass illustrate the problems small users face. The PEACESAT network was founded at the University of Hawaii in 1971 and linked universities and development centers throughout the Pacific using the "experimental" ATS-1 satellite operated by the National Aeronautics and Space Administration until ATS-1 finally drifted out of orbit in 1985. By that time, most of the island nations were linked to each other and the rest of the world via the commercial INTELSAT system. However, the PEACESAT members were unable to afford to use INTELSAT. They, therefore, searched for funds and a "free" satellite to reestablish their network. In 1987, the U.S. National Telecommunications and Information Administration received a congressional appropriation to restore the PEACESAT network using the GOES-3 satellite (Geostationary Operating Environmental Satellite), a meteorological satellite operated by the U.S. National Oceanic and Atmospheric Administration (Cooperman et al. 1991). Thus, although commercial satellite service is now available throughout the Pacific, PEACESAT, with U.S. government support, turned to another stop-gap experimental satellite because it was free.[2]

Another nonprofit development organization has gone even further and launched its own satellite for medical communications in the developing world. SatelLife launched a microsatellite in July 1991 that will provide store-and-forward data communications to small terminals in developing countries. SatelLife was founded by the International Physicians for the Prevention of Nuclear War to reflect their belief that the greatest threat to our common humanity is the gap that exists between health conditions in the developing world and those in industrialized countries.

Why did physicians feel compelled to raise funds for their own satellite? Because, despite modern technology, telecommunications facilities in the poorest regions were either unavailable or unaffordable. "The need in Africa for electronic mail not dependent on traditional communications infrastructures is desperate: In Zambia, international calls are billed at US$6 per minute. In Kenya, a fax costs $7.70 per page outgoing. In Tanzania . . . the minimal cost of a telex [is] a little more than US$25" (Clements 1991). African researchers will be able to use the satellite for free for the first three years and will gain access to medical libraries and other sources of expertise (Johnson 1991).

These applications may seem rather inconsequential in terms of usage and revenues lost to carriers, but commercial users are turning to bypass on a much bigger scale. In Latin America, banks, brokerages, and oil companies, among others, are establishing their own private networks on PanAmSat. Businesses in Asia now have the same opportunity using Asiasat, a satellite operated by interests based in Hong Kong. Again, commercial users choose to bypass the national carriers to get the services they need and to obtain better prices.

In most cases, users would rather not have to develop expertise in telecommunications and set up their own networks. Physicians, educators, and bankers simply are looking for affordable and reliable service. They would invariably rather deal with the carriers and leave the technical details to them, but out of frustration, and sometimes desperation, they have turned to setting up their own networks.

IMPLICATIONS FOR PRICING OF SERVICES

From the users' point of view, access to technology involves not only availability of equipment but also affordability. From the carriers' point of view, providing access to users is not cost free. They must determine how much to charge for use of their facilities. If they decide to offer the services at reduced prices or for free, they must justify this decision.

Carriers may disparage bypass as, at best, unfair and, at worst, cream-skimming their most lucrative business traffic. However, as noted above, most users do not want to bother with setting up their own networks. Rather than seeing bypass as the users' revenge, carriers can help to make it unnecessary by offering services users want and pricing that is both fair and affordable. The following are some pricing options now being offered in some industrialized countries as a result of user pressure that could also be implemented in developing countries.

Discounts

Carriers may use existing tariffs, which typically include reduced rates for high volume or off-peak use, and may provide discounts for government or other nonprofit users.

Reduced Rates for Spare Capacity

Some carriers offer reduced rates for capacity that would otherwise be unused. For example, some telephone companies offer "dark fiber" rates for access to optical fiber capacity that would otherwise be unused ("dark"). INTELSAT offers reduced rates for use of transponders on its backup satellites; this service is described as "preemptible," although it has never actually had to be preempted.

Free Access

Carriers may decide to provide access to their facilities without charge. Often, this approach is used for a trial or pilot project of fixed duration. At the end of the period, the user must pay to continue to use the capacity. INTELSAT adopted this approach for its Project SHARE and Project ACCESS, which offered free use of INTELSAT for health, education, and other development applications for a limited period. (The University of the South Pacific used Project SHARE as a transition to help them move from ATS-1 to INTELSAT.)

Reduced Prices

To demonstrate their commitment to the community or region and/or to showcase their technology, carriers may agree to offer educators or other public service users access to their facilities at reduced rates. They may obtain a waiver from an existing tariff or propose a special tariff, for example, for educational applications.

"Virtual" Networks/Volume Discounts

"Virtual" networks are available when needed without requiring the customer to pay a full leased channel price. For example, the telephone carrier may agree to make the channel capacity for video distance education available whenever needed but reserve the right to sell any unused capacity to other customers.

Fractional Tariffs

Some carriers in North America offer "fractional T1" tariffs for channel capacities between 56 kilobits and T1 rates (1.544 megabits).

Dedicated versus Metered Use

Usage-sensitive tariffs for video or high speed data channels are not yet common in the industry but could be offered if requested by users. Switched 56 kilobit capacity for compressed video is the type of "metered" use tariff most commonly available today.[3]

These options are not yet universally available, even in industrialized countries. However, they show how creative pricing can be offered in response to user demand. Users in developing countries will also have to work with their carriers to obtain flexible pricing.

IMPLICATIONS FOR TELECOMMUNICATIONS POLICY

To increase access to voice and data communications in developing countries, it is necessary to get rid of bottlenecks when possible and to create incentives for carriers to meet customers' needs and to raise capital for expansion. Two major strategies are required. The first involves restructuring to introduce autonomy and privatization. The second involves liberalization, or introduction of competition in various facilities and services.

The following are options and strategies to restructure the telecommunications sector. The critical elements include full internal autonomy, to raise capital, manage operations, hire and dismiss personnel; competition, in those parts of the network where the benefits are clear in terms of price, diversity, and quality; and oversight, to ensure that quality and standards are maintained, rates are reasonable, and priorities match national goals.

Management Autonomy

The first strategy for creating incentives to improve efficiency and innovation in the telecommunications sector is to create an autonomous organization operated on business principles. The end goal in many cases may be privatization as part of a national strategy to turn government-operated enterprises over to the private sector or simply a means of freeing the sector to raise its own capital and to introduce management policies that are rarely tolerated in a public enterprise.

However, in countries with a PTT structure, there may be several steps on the path to privatization. An intermediate phase of autonomous public ownership may be required as a result of pressure by influential public entities and/or unions. British Telecom, for example, went through this process. Several developing countries have chosen this model, including Singapore, Nigeria, Fiji, and Mexico.

Others see the autonomous government-owned corporation as simply an intermediate step. Malaysia intends to fully privatize the autonomous corporation it established in 1987. Sri Lanka has also established a Telecommunications Corporation as an intermediate step toward privatization.

The difficulty with a government-owned corporation is that it still may be subject to many political and bureaucratic constrains even if granted considerable autonomy. Malaysian telecommunications still suffer from government and union influences that reduce incentives. India has established a government corporation to serve Delhi and Bombay, and, although service quality has improved substantially, the government still dictates many policies that prevent the corporation from acting efficiently to meet customer needs.

Mixed Public/Private Corporations

Some countries have gone a step further by including private investment, often through foreign carriers. Several Caribbean and South Pacific nations operate their international telecommunications as joint ventures with Cable and Wireless. Belize operates its international facilities with the participation of British Telecom.

Another approach is subscriber investment through share capital. Brazilian subscribers, for example, must become shareholders in their regional telephone companies. This approach has raised much of the capital needed to construct the network. However, the experience in Brazil shows that mandatory subscriber investment alone is not a solution. First, Brazilian shareholders have no say in the policies and priorities of the telephone companies as shareholders would (if organized) in a regular corporation. Second, the government has not allowed the regional companies to function autonomously but instead has impounded their revenues and allocated annual funds rather than letting them reinvest their own profits. The result has been deterioration of the plant in the highly populated areas and very limited expansion in rural areas. However, the new president has promised major changes, including some privatization.

Fully Private Corporations

At the far end of the continuum are countries that have fully privatized their public network carriers. British Telecom is an often-cited example, as the corporation was privatized through the issuance of shares in 1984. New Zealand has also followed suit by allowing investment by foreign carriers. In the developing world, Chile has privatized its phone system. The Philippines has several private telephone companies but has poor coordination within the sector, which is dominated by the publicly owned Philippines Long Distance Telephone Company.

In general, private corporations offer the greatest commercial benefits in terms of efficiency, competitiveness, and capital acquisition. They also are leaders in technology development and innovation in equipment and services. It should be noted, however, that restructuring and/or privatizing telecommunications may not achieve the desired objectives without some form of regulatory oversight.

Introducing Competition *liberalization*

Privatization is increasingly coupled with liberalization, that is, the introduction of competition. Developing countries are generally reluctant to introduce public network competition because they believe that there are significant economies of scale and, therefore, duplication would be wasteful. However, most agree that competition can be successfully introduced in customer premises equipment and value added networks or services.

However, lack of capacity is a serious problem in many developing countries and cannot be solved simply by selling telephones and other equipment on the open market. Several strategies may be used to provide capacity when direct competition is considered unacceptable. The first is resale, in which third parties may be permitted to lease capacity in bulk and resell it in units of bandwidth and/or time appropriate for business customers and other major users. This approach may be suitable where some excess network capacity exists (for example, between major cities or on domestic or regional satellites). The second is new services by franchise or competition. The introduction of a new service may be accelerated by issuing licenses for franchises. This approach has been used for cellular radio in Argentina and Mexico, for example. It allows foreign investors with the necessary capital and expertise to provide the service more quickly than it could be offered through the PTT.

Satellite services such as data communications also may be offered through one or more private licensed carriers. Again, this approach is likely to get service installed much more quickly than through the PTT. For example, private banking networks using VSATs have now been authorized in Brazil.

Both of these technologies may be considered a form of bypass, although they may introduce services previously unavailable. The country must weigh whether the advantages of licensing the service to private carriers outweigh any dangers of cream skimming. An advantage of both these technologies is that they provide services that were not previously available to important consumer groups. For example, cellular radio may be used for fixed communications to provide public call offices and community telephones where they were not previously available, as well as to provide service to more affluent drivers. VSATs may provide important data links for important national industries such as petroleum and banking.

However, there are dangers of distortion if policy makers do not carefully frame incentives. In Malaysia, the price of installing a cellular radio is 1.5 times the national per capita GNP, thus excluding rural communities that were also supposed to benefit from the service (Wellenius et al. 1989). Policies may be framed to generate funds for other priorities. For example, carriers may charge license fees and/or operate base stations or uplinks for the private networks. They might also require that a percentage of the revenues of these services be allocated to implement other services, such as low cost fixed cellular outlets and thin route telephony via satellite.

Local Companies. Although in most countries there is a single carrier that provides both local and long-distance services, it may make sense to delineate territories that can be served by local entities. In the United States, the model of rural cooperatives fostered through the Rural Electrification Administration has been used to bring telephone service to areas ignored by the large carriers. Local enterprises are likely to be more responsive to local needs, whether they be urban or rural. An example of this approach in urban areas is India's Metropolitan Telephone Corporation, established to serve Bombay and Delhi. As noted above, service quality has improved substantially, although limited autonomy from the government has hindered the pace and scope of innovation. Local companies also provide telephone service in Colombia. Cooperatives have been introduced in Hungary. A disadvantage of this approach is the need for local expertise to operate the system, which is likely to be in particularly short supply in many developing countries.

Franchises for Unserved Areas. Another approach to serving presently unserved areas is to open them up to private franchises. Large carriers may determine that some rural areas are too unprofitable to serve in the near term. However, this conclusion may be based on assumptions about the cost of technologies and implementation that could be inappropriate. A more-innovative carrier willing to use more-appropriate technology such as rural radio systems or small earth stations and to cut costs by hiring local people rather than bringing in outside staff may be able to run the system at a profit. In Alaska, small telephone companies operate local service in villages where earth stations were installed by Alascom, the statewide long-distance carrier. Some U.S. states are considering competitive franchising of territory that the large carriers claim is prohibitively expensive to serve (Parker & Hudson 1992).

The Fallacy of Deregulation

The implicit assumption of the PTT model appears to be "What is good for the PTT is good for the country." Stated another way, because the PTT is part of the government and the government exists to serve the needs of the public, the PTT must, therefore, be serving the public interest. An implied corollary is that there is no need for participation by users, and besides they would not have the expertise to make any informed contribution. Yet, as discussed above, user participation is critical if appropriate new technologies and services are to be introduced and rates are to be kept affordable.

The changing telecommunications environment is often referred to as "deregulation." In fact, virtually everywhere but the United States and Canada, the result of changing the telecommunications structure has been the need to introduce regulation where none existed. Under the PTT model, there is no separate regulatory function: decisions on frequency allocation, standards, and prices are typically made by the PTT or a related arm of the government.

When an autonomous structure is established, there becomes a need for oversight. This is particularly true in a monopoly environment to ensure that the single provider does not abuse monopoly power in setting rates or discriminating among customers. As Goldschmidt points out, liberalization of itself cannot assure competition-absent structural and behavioral restraints on former, or "transitional," monopolists (Goldschmidt 1991). However, even a competitive environment is likely to require some oversight to ensure that service providers are competing fairly and that their collective activity is serving the national interest.

The functions that a regulator will need to perform, or at least supervise, include:

1. Pricing. Where competition exists, the marketplace may serve as a means of ensuring that prices are reasonable. However, for monopoly services or where demand far exceeds capacity, regulatory intervention may be necessary. Typically, the concern is that prices will be set too high, either to obtain excess profits or to ration scarce capacity. High prices may also conflict with national goals of providing affordable access to services.

   However, oversight may also be needed to ensure that prices are not set too low! Government pressure not to raise rates may leave the carrier without adequate resources to maintain and upgrade facilities. For example, some U.S. rural cooperatives face severe pressure from their members to maintain very low rates. They have found that it is important to have arm's length oversight to ensure that rates are not set too low.

2. Explicit subsidies. Internal cross-subsidies are often used to provide affordable services in less profitable areas. With the introduction of competition, it is important to make such subsidies explicit where they remain necessary so that a

dominant carrier cannot use them to engage in predatory pricing.

3. Quality of service. With deregulation comes the potential for degradation of service in less profitable areas as carriers strive to expand and upgrade the most profitable services. Concern about the deterioration of rural services has been expressed at the state level in the United States as well as by consumers in Canada, the United Kingdom, and Australia. Monitoring of service quality is likely to be important in developing countries, where one of the main objectives of restructuring the sector is to improve quality of service. It may also be possible to tie service quality standards to other incentives, such as pricing flexibility.

4. Criteria for access by multiple providers. Where service competition is allowed, oversight will be required to ensure that all have equal access to the network and that the provider of facilities does not enjoy an unfair advantage if it also offers services.

5. Standards. Uniform standards are required to ensure that equipment is compatible and of acceptable quality. An impartial standards agency will ensure that a dominant carrier or supplier does not introduce standards that unreasonably discriminate against other vendors.

CONCLUSION

The goal of using telecommunications to contribute to national development requires an active government policy to ensure that telecommunications plans and services are designed to meet national goals. It also requires flexibility and innovation in services, equipment, and pricing to respond to user needs. New technologies offer new opportunities but they also pose new challenges for telecommunications carriers. The most important challenge will be to plan and manage telecommunications not only as a source of revenue but also as a strategic resource for development.

NOTES

1. The author was a special advisor to the Maitland Commission and drafted sections of the report on the role of telecommunications in socioeconomic development.

2. Some PEACESAT members went back to using HF radio nearly 20 years after ATS-1 demonstrated that satellites could reliably and affordably replace HF radio in Alaska and the Pacific!

3. For more information on these and other pricing options, see Parker, Edwin B., and Heather E. Hudson, *Electronic Byways: State Telecommunications Policies for Rural Development.* (Boulder, Colo.: Westview, 1992).

REFERENCES

AT&T. *The World's Telephones*. Bedminster, N.J.: AT&T, 1990.

Blair, Michael L. "VSAT Systems in Developing Countries." *Proceedings of the Pacific Telecommunications Conference*, Honolulu, February 1988.

Clements, Charles. "HealthNet." Cambridge, Mass.: SatelLife, May 1991.

Cooperman, William, Lori Mukaida, and Donald M. Topping. "The Return of PEACESAT." *Proceedings of the Pacific Telecommunications Conference*, Honolulu, January 1991.

Goldschmidt, Douglas. "On Curbing Predation in Newly Competitive Markets." Unpublished paper, Greenwich, Conn.: Alpha Lyracom (PanAmSat), 1991.

Hudson, Heather E. *Communication Satellites: Their Development and Impact*. New York: Free Press, 1990.

Hudson, Heather E. *When Telephones Reach the Village: The Role of Telecommunications in Rural Development*. Norwood, N.J.: Ablex, 1984.

Hudson, Heather E. (Ed.). "Innovative Strategies for Telecommunications Development." *Telematics and Informatics* 4 (Spring 1987): 97–108.

Hudson, Heather E., and Lynn York. "Generating Foreign Exchange in Developing Countries: The Potential of Telecommunications Investments." *Telecommunications Policy*, September 1988.

Hukill, Mark A., and Meheroo Jussawalla. "Telecommunications Policies and Markets in the ASEAN Countries." *Columbia Journal of World Business* 24 (Spring 1989): 43–56.

INTELSAT. *Annual Report*. Washington, D.C., 1990.

International Development Research Centre. *Sharing Knowledge for Development: IDRC's Information Strategy for Africa*. Ottawa: International Development Research Centre, 1989.

International Telecommunication Union. *Information, Telecommunications, and Development*. Geneva: International Telecommunication Union, 1986.

International Commission for Worldwide Telecommunications Development (The Maitland Commission). *The Missing Link*. Geneva: International Telecommunication Union, 1985.

Johnson, Tony. "Microsatellite That Turns Information into Medical Power." *The (Manchester) Guardian*, April 26, 1991.

Mayo, John K., G. R. Heald, S. J. Klees, and M. Cruz de Yanes. *Peru Rural Communication Services Project Final Evaluation Report*. Washington, D.C.: Academy for International Development, 1987.

National Research Council, Board on Science and Technology for International Development. *Science and Technology Information Services and Systems in Africa*. Washington, D.C.: National Academy Press, 1990.

Parker, Edwin B. "MicroEarth Station Satellite Networks and Economic Development." *Telematics and Informatics* 4 (1987): 109–12.

Parker, Edwin B., and Heather E. Hudson. *Electronic Byways: Telecommunications Policies for Rural Development*. Boulder, Colo.: Westview 1992.

Parker, Edwin B., Heather E. Hudson, Don A. Dillman, and Andrew D. Roscoe. *Rural America in the Information Age: Telecommunications Policy for Rural Development*. Lanham, Md.: University Press of America, 1989.

Pierce, William B., and Nicolas Jequier. *Telecommunications for Development*. Geneva: International Telecommunication Union, 1983.

Saunders, Robert, Jeremy Warford, and Bjorn Wellenius. *Telecommunications and Economic Development*. Baltimore: Johns Hopkins University Press, 1983.

Tietjen, Karen. *AID Rural Satellite Program: An Overview*. Washington, D.C.: Academy for Educational Development, 1989.

Wellenius, Bjorn, Peter A. Stern, Timothy E. Nulty, and Richard J. Stern (Eds.). *Restructuring and Managing the Telecommunications Sector*. Washington, D.C.: World Bank, 1989.

13

Agenda for the 1990s

Klaus Grewlich

COMMUNICATIONS TECHNOLOGIES — AN AMAZING RACE

The Power of Information and Technology

Information and knowledge have always conferred power on those who have it and know how to use it. The dissemination of information and technology — in both the civil and military fields — can be a precursor to political shifts or consolidation in the national, regional, and global political structures as well as economic power structures.

The information revolution is changing the international economic system, transforming national, political, and business institutions, and, thus, is affecting not only the nature of national sovereignty but also the relative strength of the leading nations and economic actors in world affairs, in the same way that the long-range gunned sailing ship or the development of steam power massively increased the relative power of certain nations and decreased the relative power of others.

The significance of technology as an outstanding factor of the world economy is not new. World history always has also been economic and technological history. What *is* new is the acceleration of technological developments, the super-fast sequence of generations of technology, and the fierceness of competition in an international race requiring an all-out effort.

The intelligent use of "information," like energy and transport, is a basic requirement for economic success. One need only think of the worldwide network of reservation systems for civil aviation, hotels, and car rentals. International credit card systems could not exist without instantaneous access to verification centers via data communication networks. Numerous data bank services enable users to quickly receive information from a wide variety of data bases, which must be constantly updated. (An example of this is the financial information systems.) In industry, "just-in-time" production, with is considerable savings in logistics and warehousing costs, is made possible by electronic data flows. In trade, electronic data flows, resulting from efforts to standardize shipping documents, offer considerable cost-saving potential. Major banks transfer money around the globe in a matter of seconds: Put simply, money becomes electronic information and bank transactions are telecommunication. Information and communication have become an autonomous, globally active production factor. The world's financial marketplace will never recede to its old national constraints. On the basis of modern telecommunications, ideas and money move across borders at a speed never before seen.

The basis of the information revolution is the programmable computer. The integration of computer technology and telecommunications is now enabling us to fully exploit the potential of electronic data processing, just as the construction of a railway network in the nineteenth century mobilized the steam engine and led to a period of enormous economic growth. The overcoming of time and space through advanced telecommunications (in other words, the potential ability to have access to electronically transmitted information at every point on the earth simultaneously, to produce information as an economic or cultural asset, and to transmit it instantaneously across the country or across the continent, thereby creating a state of "simultaneity" and "omnipresence") is opening the way for a new stage of economic and cultural development.

The new production factor, "information," is triggering dynamic concentration, restructuring, and diversification processes and new alliances in the economy. Some manufacturers of communications equipment are moving into the computer field, that is, microelectronics; computer manufacturers are producing telecommunications equipment or establishing varying degrees of cooperation. The race for global networks is a secular phenomenon. Innovative publishers are looking toward multimedia products so that books no longer appear simply in print form but are also filmed for television (future high quality TV) and video and made available as abstracts in data banks.

The "spirit of 1993" will lead to European and global cooperation schemes. The current high-tech revolution seems sometimes like a poker game in which the corporations that are playing raise their stakes with ever higher investments in order to stay in and benefit from the eventual bonanza. Increasingly, the solution is seen in the establishment of cross-border alliances.

The relative position of industrial nations in this amazing race is determined not only by the economic-technological potential but also by the intelligent use of framework conditions such as "deregulation and regulation," that is, the legal framework for technological and economic activities, including open standardization policies for networks, equipment, and services.

Such efforts toward liberalization and harmonization in the field of telecommunications are beneficial for the world economy because they remove obstacles for trade in goods and services by creating more liberal framework conditions. At the same time these endeavors show that the race has begun for "prime locations" in international data communication and telecommunication network based services — locations that offer a first-class telecommunications network, few restrictions on use of the network, cost oriented fees, and an open attitude toward service based businesses. In the future, such locations will become the centers of the global information and communication economy. Therefore, many observers consider the right mix of deregulation and liberalizing efforts to be a crucial device to achieve improved competitiveness.

Worldwide, there is likely to be important growth, particularly in the field of telecommunications. Sales in this area are expected to increase by an average of approximately 8 percent annually over the next ten years. Thus, telecommunications might increase from approximately 2 percent to 7 percent of GNP or more in most countries of the Organization for Economic Cooperation and Development (OECD) by the end of the twentieth century. The number of telephones and hand-held sets to be sold in the 12 countries of the European Community (EC) is estimated to increase by about 80 percent between 1990 and 1997, with annual spending rising from about $20 billion to about $35 billion. Annual growth rates for computer hardware are expected to be impressive, and the growth rates for the online and global network markets are predicted to be as much as 20 percent. Particularly dynamic growth is expected in the software market, in which experts predict annual growth of as much as 25 percent worldwide, with considerable potential for small and medium sized businesses. Growth rates of 20 percent to 30 percent for the field of value-added services toward the end of the 1990s have been estimated at OECD.

The Global Village

If not yet a global village in the full sense, the industrialized regions of the world have certainly grown together as a result of the use of the telegraph, the telephone, and, most recently, satellite and computer based information networks.

The Commission of the EC has estimated the impact of the telecommunications sector on the Single European Market and come up with impressive forecasts. It looks like the information and telecommunications industry will overtake the automotive industry as the single most important sector of production and services. The customer will be seeking a greater diversity of products and convenient full-service handling, including one-stop shopping and itemized billing, to name just a few service features.

The trend toward free trade in and across national markets — notwithstanding the difficult situation in the General Agreement on Tariffs and Trade (GATT) — will gain further momentum and will increase the pressure to open monopoly services to competition.

The last decade of this century will see the realization of technological innovations that were initiated in the 1980s. Digitalization will be completed, and optical fiber technology and the miniaturization of electronic components will progress further. High definition television (HDTV) may eventually become more of a household convenience than an expert concept for business and institutional customers.

Technologically speaking, the 1990s may be the decade of full national and international integrated services digital network (ISDN) implementation with the enhanced added value service capabilities of a high-speed, broadband network. Sophisticated software for network management and for user equipment will be the key factor for technological innovation.

By the end of the twentieth century, telecommunications, thus, will be more visual, more intelligent, and more customized to the personal requirements of individuals. Mobile communications, in particular, will become increasingly important to users on the move, who will need ubiquitous access to voice, images, text, and data to remain competitive. All these requirements will increase the pressure to standardize norms and to realize economics of scale by international cooperation and joint ventures.

In an environment of rapid change and shortened life cycles of products, long-term strategic business planning is inherently risky. However, above all, the pace of innovations offers tremendous opportunities.

We may see an extraordinarily dynamic period of change — change driven by customer demand for a wider range of services, technological innovation, and, hopefully, liberalized trade policies.

The emerging global business environment will be characterized by both conflict and more competition but also by greater cooperation.

The World Market

In spite of the extraordinary dynamic expansion of the world market for information and communication technology from an estimated $500 billion to $1 trillion by the end of the 1990s, some of those who are now involved in the race will be forced out of it. The enormously investment-intensive sequence of technology generations that succeed each other at a dizzying speed and, above all, the "system character" of information and communications technology will in all likelihood lead to a displacement competition. There will be winners and there will be losers, or to put it more cautiously and thus avoid the impression of a going-for-broke game, some countries and regions will win more than others.

In view of this perceived development, many countries are today seeking to consolidate or improve their level of technological performance, although by various means and to varying degrees. As a result, power politics could affect market mechanisms and bring states into direct confrontation with one another.

Concern that foreign trade conflict potentials will become more evident cannot be brushed aside: issues such as subventionism and export promotion; export controls; trade restrictions (using partly the "infant industry" argument); arrangements such as "semiconductor agreements"; government induced technology policies; so-called precompetitive cooperation in the field of integrated circuits; "telecommunications trade acts" with "priority lists," distinguishing between friendly and unfriendly actions; new requests for "national content and origin" in the field of chip trade; programs for the creation of secure standards for commercial data networks, excluding nonnational companies; proposals concerning quota systems in the field of media services; "competitiveness acts" designed to improve the performance of national industries all may add to accelerating the global race and increase the danger of political frictions in the field of information and communications technology. At the same time, these issues clearly show that cooperative political solutions are vitally needed.

The situation is even more complex: some political quarters fear that transnational companies or enterprises based in the main technology centers of the world will make the states the instruments of their industrial strategies and economic or technological power struggles, because despite dynamic growth rates in the transnational information and communication economy, excess capacities may be generated and there may not be room for all those who are presently in the running. For the telecommunication equipment companies, for instance, the financial squeeze is severe: a minimum market of $14 billion in business over ten years is needed to cover the cost of developing the new generations of computerized digital telephone exchanges, but in Europe, for instance, no single national market is that big.

This might lead to the shakeout announced for telecommunications equipment manufacturers. A similar shakeout might take place in the field of the operators. There is, thus, a clear argument for establishing alliances to cope with increasingly sophisticated user requirements, increasing high tech investment, and shortened product life cycles. Such alliances or transnationally linked companies may, in many cases, have a stabilizing effect on international economic and political relations.

COOPERATIVE COMMUNICATIONS POLICIES

It will not be possible for one country, region, or economic actor to derive the benefit of the process toward the "information society" unless it has the cooperation of others. This simple and central fact has to be kept constantly in view. The international law of the past, which was intended to define areas of sovereignty of different states, was simply a law of international coexistence. It is increasingly being replaced by a more highly developed law of international cooperation. This more enlightened law asks the states to cooperate positively at the bilateral, regional, and global levels in an ever increasing number of fields, or at least, when adopting measures that will affect third parties, to consider the latters' interests and to minimize the negative consequences as much as possible. This spirit is the foundation of international organizations like GATT, International Telecommunication Union (ITU), and OECD.

Will it be possible to find political problem-solving mechanisms for the critical areas of the transnational information economy, particularly transborder telecommunications/information services, before attempts by individual countries to protect their own interests and establish regulations create obstacles that will be difficult to overcome? It is clear that no individual country or region can deal with this task alone. Basically, this is a task for the world community. If we adopt these aims, cooperative communications policies within the established international organizations, for example, GATT and the ITU, can play a decisive part. The skeptics will shake their heads and point to the divergent policies, varying political priorities and (not least) the considerable differences in levels of technological achievement within the community of nations. Nevertheless, I believe that cooperative communications policies must be tackled and are feasible. The following are several promising indicators and objective constraints and opportunities in support of this theory.

Standardization

Standards can affect many aspects of telecommunications, such as interconnection of networks, interworking of terminals, competition in the marketing of equipment and services, and, notably, speed in the adoption of new technology.

In information technology and telecommunications, standardization implies the compatibility of systems, which is vital to the promotion of interconnection and the interworking of an increasingly complex and varied system. What matters here is the combination of, on the one hand, deregulation (including the opening up of public procurement) and, on the other hand, the provision of common open standards that should have the effects of reducing concentration and increasing competition. Common open standards means "manufacturer neutral" and not "de facto" international standards! These open neutral standards should also be especially advantageous to small and medium sized enterprises who wish to supply equipment.

Where the advantage of economies of scale are great because of considerably reduced research and development (R&D) costs, the existence of open standards should ensure that monopoly suppliers of national telecommunications administrations will, in the future, have to compete. This new development would certainly benefit consumers as lower costs are passed on but should, ideally, also be a strategic advantage for such firms on international markets.

Mutual recognition of standards and respect for the principle of subsidiarity are enlightened policy objectives, but they do not necessarily guarantee the interoperability of the telecommunications networks and services. Among the measures to be introduced to open the markets and to ensure free movement of goods and services, technical standardization, notably in the field of telecommunications, occupies a place of prime importance.

It has been estimated that — Europewide — standards could generate annual savings of 5–10 percent in the equipment market alone, corresponding to some 1 billion European Currency Units per annum, or approximately 50 percent of the 2 billion that the European telecommunications industry was spending annually on R&D in the early 1980s.

In Europe, the following mechanisms are available for setting standards: the EC legislation, the formal standardization machinery, and the initiation of innovative projects, including (to some extent) the provision of support funds.

Standardization mechanisms in Europe have sometimes caused concern outside Europe for two particular reasons: first, it was thought that European standards would be a "defense line" built to prevent intruders entering the "magic single market," and second, it was thought that strong European standards institutes would compete with the worldwide standards organizations for expertise and money and that this competition would be prejudicial to these organizations and consequently to world trade.

The Commission of the European Community has given answers to these suspicions. First, probably 80–90 percent of all European standards produced so far are nothing but international standards implemented in a harmonized way and given a higher legal status. The remaining European standards are a replacement for a multitude of conflicting national standards and are, thus, only streamlining them for the benefit of everybody. Second, the European Telecommunications Standards Institute (ETSI) represents the focal point of European activities in telecommunications standards, and it has been structured in a way that shows the intention to be open and transparent, quite the opposite to building a "fortress." Evidence of this is that ETSI is also open to non-European organizations concerned with telecommunications, who are invited to participate in meetings of the Technical Assembly. Third, concerning the negative effects that an increase of European standards activities would have on international organizations, it has not been demonstrated that the contributions of Europe to these institutes have diminished. In fact, an increased number of proposals have served to enhance the importance of international standardization.

Again, however, why European standards? This question can be answered from one of two directions. The first is why not national standards? The second is why not worldwide standards? It is legitimate and intelligent to use the political drive toward completion of the internal market by 1993 also in the field of telecommunications and to overcome purely national terms of standardization. Also, a worldwide standard is certainly better than a purely European one. This is indeed the position of ETSI. A worldwide standard is seen as the ultimate objective, but the larger the number of partners needed for global consensus, the more divergent are the interests that have to be recognized, the longer the path, and the more likely the fact that markets somewhere cannot wait. Standardization is as much about timing as about content.

As to the worldwide standardization organizations, the ITU's Consultative Committee for International Telegraphy and Telephone (CCITT) drafts recommendations making communication between public telecommunications services possible. Despite its antiquated title, the CCITT's area of activity includes modern technology, such as ISDN. The ITU's aim is worldwide standardization. Until now, the typical times for ITU standardization have been two years for a standard following a wholly existing pattern, five years for a standard that was largely new, and ten years when the area was so new that a framework for the norms had to be established first. These periods will have to be compressed!

A vital question is whether cooperation between the emerging regional standardization bodies such as T 1, ETSI, and respective organizations in the Pacific rim may lead to ITU's role is being affected. Such a perception should, however, not cause unnecessary dramatization. ITU remains the internationally recognized standardization organization, with a reputation earned with its constant, persistent, and nonpartisan work in its standardization bodies. One of ITU's outstanding accomplishments in terms of international policy and economic public international law is its having elevated "interconnectivity" — that is, the principle that anyone in the world can exchange data, pictures, and language with anyone else on the basis of compatible networks and telecommunications services — to an international legal standard. This accomplishment has two components: on the one hand, a legal framework for the new electronic services was increased that does not retard innovation; on the other hand, the technologically less developed countries are protected from being cut off from the modern telecommunications system. This is an especially impressive manifestation of the principle of "universality" in global telecommunications, which is a major aspect of cooperative communications policies.

Instrumental for this achievement and another promising sign that cooperative communications policies are possible was the outcome of the World Administrative Telephone and Telegraph Conference, which met in December 1988 in Melbourne and produced a binding international instrument. This agreement contributes to an interdependent, open world in the spirit of cooperative communications policies and increases the effectiveness and legal consistency of the international economic system, particularly in the areas of trade and direct investments, fiscal and monetary relations, and cultural exchange.

Expansion

Because telecommunication is, indeed, the fuel for progress, we cannot afford to cut off one-half of the globe from this resource. The fact that three-fourths of the world's telephones are installed in only nine countries is a potentially dangerous barrier to greater self-determination and more social justice in developing countries.

It is quite clear that the leading nations in telecommunications must increase efforts to provide the developing nations with affordable equipment and services. The 1989 Plenipotentiary Conference of the ITU in Nice made a substantial contribution to solving complex development problems. Although the planned structural reform of the ITU has been postponed and too little attention may have been paid to the interests of users and innovative newcomers selling communication services, a promising breakthrough in international telecommunications policy has been made in the form of ITU's additional work in the field of technical cooperation — the establishment of a new bureau. A "high level group" promotes this matter further and will soon come to concrete results, strengthening the cooperative framework for communications policies.

Ecology

Telecommunications is a genuinely ecologically friendly industry. In bridging time and space and facilitating the basic human need to communicate, telecommunications technology conserves resources and energy. In the long run, this truly protects our environment.

Nevertheless, we have only begun to exploit the potential of telecommunications in, for example, monitoring the state of the rain forests or the extent of environmental pollution. Telecommunications can — if the proper cooperative framework is established — make significant contributions to easing the traffic congestion in our cities and more effectively channeling energy consumption.

Flow of Information

With regard to international media policy, which features prominently in the United Nations Educational, Social and Cultural Organization, it might be necessary in the years ahead to adhere even more to the principle of the free flow of information, to further objectify unnecessary disputes, and to develop an international media policy that, on the basis of practical media assistance with the aim (among others) of establishing independent news agencies and television production capacities in the Third World, would reduce actual or potential imbalances in flows of information, to the extent that this implies not the control or constraint of such flows but their release and expansion; not national isolation but international competition; not monopolies by the big countries but help for the small ones to be able to compete; and not government monopolies of information media but pluralistic diversity. The "International Programme for Development of Communications" is a path we should continue along, fully in keeping with cooperative communications policies.

High Definition Television

In the last decade of this century, there will be a tremendous struggle over standards and markets for HDTV. The issue is one of "global telepresence." It involves technology and markets worth hundreds of billions of dollars. It concerns the replacement of some

600 million television sets in the next decade and the production of new programs.

However, we should not examine this question from the aspect of possible distribution struggles but in terms of the opportunity for cooperative communications policies. The electronic media transcend space and time. As a result of worldwide telecommunication and the global audiovisual medium of television, dream and reality, technology, and human consciousness are being transformed into new systems and global interrelationships. At the same time, a uniform television norm is aimed at preserving, enhancing, and broadcasting the cultural attractiveness of individual regions.

The three-cornered interrelationship of progressive microelectronic components, consumer electronics, and distributive technology, as well as the market strategies and struggles for technological dominance associated with it, may, it is true, lead to a fragmentation of the world of television of the future, but such a development would militate against the universal character of modern telecommunication and the media. It is a major political task to find evolutionary and compatible solutions that will be acceptable to all members of the international community and secure the foundations for cultural exchange.

Legalities and Economics

Free flow of information does not mean that information across borders will be without cost. We have yet to find the adequate framework conditions. The need for free and ubiquitous access to more and more information is raising awareness of justified concerns.

Specifically, the following aspects are involved: data protection, copyright protection, legal aspects of competition, computer criminality, the regime for communication-based services (that is, the applicability of GATT rules for trade with goods to the trade in services), the definition of "value-added services," and an attempt to develop a regulatory framework for the "new telecommunications order" as well as rules under international economic law for conflicting technological and industrial policies. Because the sector of information and communication technology is a prospective one, the following considerations will focus not only on the lex lata but also on the lex ferenda, taking into account the present status of GATT negotiations.

With regard to the protection of personal data in transborder data flows, for instance, a balance must be found between the need to protect the privacy of individual citizens and the importance of the free flow of data across borders for international business and science. The remote processing of data abroad would also fall into this category. At the international level, the divergent needs of data protection and free flow of data are dealt with in the OECD's "Guidelines for the Protection of Privacy and Transborder Flows of Personal Data" of September 23, 1980, and the Council of Europe's "Convention for the Protection of Individuals with Regard to the Automatic Processing of Personal Data" of January 28, 1980.

In addition, new legal bases for the control of various forms of computer criminality, that is, the misuse of data processing, particularly so-called hacking, must be established. Penal codes must introduce new criminal offenses such as computer fraud, forgery of data used in evidence, spying out of data, alteration of data, and computer sabotage. If access is gained to commercial or industrial secrets, the law on unfair competitive practices may also come into play. The next step would be to harmonize national penal provisions concerning computer criminality. In international institutions, an exchange of pertinent information has begun and has led to the development of a typology of computer offenses. True harmonization, however, is still far from being achieved.

Important, too, is a set of international rules for the protection of data bank searches. By analyzing data bank searches (for instance, those undertaken by industrial companies), conclusions can be drawn as to the company's technical performance and major areas of R&D. Such information is of great interest to competitors. Although providers of data bank services give the assurance that the use of the data banks remains absolutely confidential, international legislation guaranteeing such protection is desirable. At the same time, technical safeguards to prevent unauthorized access to data should be reinforced. Some measures particularly relevant to data and communications security in the sense of network integrity/data security are authentication to prevent unauthorized access to data bases, cryptological safeguards, and radiation shielding. International agreement concerning the "electronic signature" should be reached as soon as possible.

On April 11, 1985, the OECD ministers adopted a "data declaration" regarding commercial data flows. In this document they have, in particular, announced their intention to keep access to data banks and information open (this must be seen also against the difficulties associated at an earlier stage with access to data banks having a dominant market position) and to undertake efforts to maintain the present state of liberalization with regard to transborder data flows and eliminate remaining obstacles. Although it is not legally binding, this OECD declaration on transborder data flows, which combines classical export liberalization with a basically new "claim" to the right of access to critical information resources, is very important for the development of cooperative communications policies.

Political Changes

In Eastern Europe and the Soviet Union, telecommunications have been instrumental in the process of political change. A recent OECD study has described communication technologies as "freedom technologies."

Now, communications are vital for sustaining the democratization process and for economic recovery. It is a major challenge to open up and to make available adequate tools to support efforts in the telecommunications development of Eastern Europe.

The tasks to be accomplished are so important that even the biggest countries and economic actors from the industrialized world may have difficulties in meeting them. Thus, cooperative policies are of vital importance.

For the economic actors such as telecommunication manufacturers and operators, an intelligent mixture of competition and cooperation — in harmony with national requirements — is needed.

There are great opportunities. The Eastern European dynamics is a challenging area for demonstrating the superiority of the market philosophy and for proactive institutional innovation in a new field for cooperative communications policies.

14

The Challenge of Change

Meheroo Jussawalla

This volume has covered the regulatory policies of the major Organization for Economic Cooperation and Development (OECD) countries as well as some parts of the developing world like Africa, Latin America, and Southeast Asia. Some of these regulatory policies are being subjected to rapid changes because of the vast growth of innovative technology and the convergence between telecommunications, computers, and satellites. Both users and technology are challenging the regulators to formulate policies that are in keeping with developments in the real world. Even after the divestiture of American Telephone & Telegraph (AT&T), policies in the United States had to be changed with regard to spectrum allocation, the challenge of integrated services digital network to the cable industry, mobile communication systems, and multimedia. Many developments are taking place in Europe with the advent of the Single Market in the arena of telecommunications equipment and services. Global markets are being supplied by new entrants taking advantage of growing competition in the industry. At the same time, regional trade blocs are being formed to meet the challenge of the single market in Europe. The General Agreement on Tariffs and Trade is unable to deal with the problems arising in trade in services, and the negotiations are being stalled even with regard to the Telecom Annex. Under the circumstances, let us examine the contributions of the various authors to show which countries are meeting these challenges with greater success than others.

A major advantage of the revolution in telecommunications technology is that different countries and their administrations have before them a wide variety of options in equipment, processes, and services from which they can find the systems that are best suited to their national requirements. The selection generally depends on political, economic, and sociological factors, but the aim of the investment policy in each country is to obtain maximum welfare at minimum cost subject to limitations of available technology, resources, market imperfections, and institutional constraints.

Regulatory theory originating from the seminal article by Averch and Johnson (1962) focused on rate-of-return regulation under capital intensive monopoly. Stigler (1971) argued that regulators often come to favor the firms they regulate, although Peltzman (1976) considered the regulatory process as a transfer of wealth that generated productivity and growth. Regulatory theory has been applied to telecommunications by Noam (1983) in an analysis of present and future regulation in the United States, and Mosco (1984) presents the international consequences of telecommunications policy. Technological change has grown at an exponential rate and become the major force in driving changes in telecommunications policy (Snow and Jussawalla 1986). In examining the impact of technological change on the industry, Lamberton (1983) shows how information economics has contributed to an understanding of the process of technological change. Likewise, Jussawalla (1985) has examined the relationship between technology and productivity increases and concluded that productivity is higher in information-related industries. Most of the literature surveyed recognizes that telecommunications and information technologies serve as catalysts of the information revolution. OECD countries enter into vigorous competition to stay at the cutting edge of technology in order to boost their exports in global markets.

The United States was probably the only country that over decades entrusted its telecommunications services to a private monopoly and regulated its operations. Most other nations resorted to statutory bodies called Post, Telephone, and Telegraph (PTT) organizations that virtually controlled fiefdoms in their regions. Their continuing intervention in the sector has grudgingly adapted itself to technological changes as seen in tariff and nontariff barriers raised against the free transborder flow of data (Jussawalla and Cheah, 1987). Many developing countries used their PTTs as sources of general revenue, and this was also the practice in Europe. Interventionist policies extended to restrictions on imports and trade regulations that interfered with technology transfer. These trends have been examined in this volume in the chapters dealing with Africa and Latin America. The pervasive changes in technology have challenged the policy makers in these countries to revise their regulations, and gradually the PTTs find compelling reasons to liberalize their operations, especially in the area of value-added services. The U.S. model was not fully adopted either in the United

Kingdom or in Japan for purposes of deregulation, chiefly because there were other considerations of ownership, labor unions, and the impact on the economies of those countries. The UK established a duopoly in order to justify private ownership of the industry, and in Japan, Nippon Telephone and Telegraph Company (NTT) was fairly slow in giving up its total control over the domestic market, as discussed by Oniki in this volume. The international services supplier in Japan with monopoly control was Kokosai Denshin Denwa (KDD), and it resisted competition for as long as it could, until foreign pressures enabled the formation of two competitors.

The European Commission in 1980 recognized the challenge to its telecommunications industry from the United States and Japan, as was shown in the trade deficit in the sector, which by 1991 had grown to $35 billion. As shown by the contributions of Chamoux and Grewlich, almost all European Community (EC) governments resisted the idea of restructuring their PTT operations to meet the needs of the Single Market. France continued with its PTT but provided sophisticated services like PBXs and videotex to its users through joint ventures or participation by the private sector. Domestic facilities were improved within the framework of government ownership. Although the Green Paper has no legal force, it focused on the goals establishing an efficient regional network and making private suppliers of telecommunications equipment more viable to compete in global markets. The most difficult problem for the European Commission is that political issues stand in the way of fully implementing the Green Paper. Complete deregulation of PTTs is not agreeable to the member states (Dizard, 1992). As rightly pointed out by Carpentier (1991), telecommunications has been one of the most contentious and sensitive issues for the EC. It became imperative for the Directorate General XIII to ensure that the private sector would get a fair chance to compete with the PTTs. In June 1990 this was started through the Open Network Provision for equal access to the government-controlled networks. Most countries were reluctant to introduce full competition for their services. The other argument was that equipment standards had not been established for harmonizing the interconnection between the different networks. The European Telecommunications Standardization Institute has been working on this issue for some years now.

Satellite communications, as we have seen in this volume, have become a difficult problem for coordination in the EC. However, a regional satellite system, Eutelsat, has been providing services in an efficient manner. The EC in 1990 proposed deregulation of control over earth stations in Europe to permit unrestricted access to satellites as is available in the United States. The member states have agreed to implement this proposal and to upgrade services for cellular telephones and mobile services.

In order to meet the competition both regionally and internationally, the industry responded by starting a number of mergers across national borders. This becomes essential in light of the estimated growth of the global market for telecommunications trade to $500 billion per year by 1995 (Dizard, 1992). The European corporations in this sector are apprehensive of aggressive competition from U.S. and Japanese competitors. International Business Machines and International Telephone and Telegraph have operated in European markets for many years, but the deregulation in the United States opened up the U.S. equipment market to European vendors. Major competitors emerged to capture the biggest market in the world by entering the United States with production facilities. Siemens of Germany, Alcatel of France, and Phillips of Holland either have established mergers with U.S. companies or are operating on their own. Cable and Wireless and British Telecom have also established themselves in U.S. markets. The latter firm owned 22 percent of McCaw Cellular, which is the largest cellular telephone operator in the United States and has recently merged with AT&T.

It is clear from this work that as deregulation has swept across the countries of the world, companies from different countries are exploiting the markets of each other in a frantic effort to capture global profits. Japan has kept its edge in the market for semiconductors, and despite the imposition of the 301 Clause of the U.S. Omnibus Trade Act of 1988, the Japanese have been supplying world markets with their electronic products. For both the Japanese and the U.S. firms, the economic stakes involved in the new European telecommunications market are sizable. Alarmed at the prospect of foreign competition, the European industry is seeking subsidies from Brussels along with policies for limiting the entry of Japanese firms into the European market. The Washington *Post* reported on August 9, 1992, that the European Council in 1989 required European origination of television programs on regional television stations in order to reduce imported U.S. programming. Despite the differences between the European, the U.S., and the Japanese systems, there are common problems that regulators face in most countries, whether developed or developing. The diversified service economy that is appearing in most countries is dependent on telecommunications technology for production and distribution processes so that consumers anywhere can get delivery at any time (Melody, 1990).

Three big changes are transforming the U.S. telecommunications market in the same way as the breakup of AT&T. One of these will challenge the monopoly of the Regional Bell Operating Companies in the local markets, because AT&T has announced its participation with a 33 percent stake in McCaw Cellular Communications, the biggest firm in the wireless market. AT&T's deal involves $3.8 billion and gives it the right to compete with the cellular subsidiaries of the regional Bell companies. AT&T will now be able to connect local customers to long-distance lines, using cellular technology. This means that AT&T no longer pays the Bell companies 50 cents out of every dollar for access to its local networks.

The second big change and challenge to the Bell companies comes from MCI, which plans to provide personal communications network (PCN) to its customers by 1994. Using a higher frequency than cellular phones, it will provide cheap pocket telephones to a mass market. By November 1992 the Federal Communications Commission (FCC) received 100 proposals for the same technology by rival firms. MCI met the challenge by announcing a national consortium for supplying PCNs. In this case it will get a large share of the video and data market and threaten the share of the Bell companies (*The Economist*, November 14, 1992). If this consortium cooperates with the cable companies in supplying television programs, the Bell companies will be further restricted in their plans for expansion.

The third big change in the telecommunications industry in the United States will come from President William Clinton, who is committed to giving the country better telecommunications. If the monopoly of the Baby Bells is opened to competition and they lag behind in high technology innovations, they may have to enter the long-distance market in order to survive. This may lead to an unbundling of the regional monopoly's services as well as the pricing of these services in order to make the pricing more cost based. These challenges will again have to be met by the regulators. Almost nine years after the regional phone companies in the United States were spun off from AT&T in 1984, these companies are feeling the pinch of enhanced competition. In 1991, Pacific Telesis made $1 billion in profit on a revenue of $10 billion (*Business Week*, October 5, 1992). Shareholders have enjoyed the benefit of these profits, but as these companies began to move out into international ventures, from Hungary to Australia to Britain, their losses are beginning to mount. They have enjoyed a monopoly that has been slow in growth because they are operating in a high technology, fast speed global network. Already the regulators are calling for ending the monopoly on local services, and the FCC is directing the Bells to allow rival carriers to use their networks to connect calls to large customers. These companies have already diversified into computers and finance software for mainframes, and U.S. West even tries real estate.

The message is clear. Monolithic local phone companies are out of sync with markets that are becoming more sophisticated and demanding customized services. New services with more usage would generate more revenue (see Table 14.1). This concept has brought rethinking of their policies to the Bell companies, and they are moving toward alliances for delivering financial data, medical

TABLE 14.1

Comparative Investments in Telephone Networks, 1990[a]

Country	Total Period Telecom. Investment (US $M)[b]	Growth in Main Lines Expansion[c] (000)	Investment in Network Expansion (%)[d]	Investment in Modernization (%)[e]	Rank
Switzerland	11,185	946	12.7	87.3	1
United States	217,509	19,398[f]	13.4	86.6	2
Sweden	9,364	896	14.4	85.6	3
Germany	76,057	7,865	15.5	84.5	4
Japan	107,251	11,335	15.9	84.1	5
Austria	8,090	912	16.9	83.1	6
Canada	28,244	3,941	20.9	79.1	7
Denmark	4,184	622	22.3	77.7	8
Norway	5,778	873	22.7	77.3	9
Ireland	2,693	420	23.4	76.6	10
Finland	5,181	842	24.4	75.6	11
Australia	16,964	2,850	25.2	74.8	12
Italy	46,785	8,249	26.5	73.5	13
Luxembourg	236	44	27.9	72.1	14
New Zealand	1,742	349	30.0	70.0	15
Spain	22,314	4,568	30.7	69.3	16
France	51,002	11,044	32.5	67.5	17
United Kingdom	33,586	7,667	34.2	65.8	18
Belgium	5,603	1,285	34.4	65.6	19
Netherlands	7,276	1,799	37.1	62.9	20
Iceland	119	37	46.8	53.2	21
Portugal	2,947	1,046	53.2	46.8	22
Singapore	1,278	458	53.8	46.2	23
Greece	2,953	1,516	77.0	23.0	24

Notes:

[a]OECD members plus Singapore.

[b]Capital investment by public telecom operators excluding land and buildings. Unadjusted for differing treatments for labor costs and CPE inclusion.

[c]Assumes cost of network modernization = total investment − cost of network expansion.

[d]Assumes an average cost of US $1,500 per new line (1989). Percentages computed using unrounded numbers.

[e]Constant 1989 U.S. dollars (adjusted for inflation and exchange rates).

[f]An increase in U.S. main lines for 1988 may be due in part to changes adopted that year in the FCC's method of collecting data in this area.

Source: OECD, adapted from *ITU Yearbook of Statistics*, MCI and US Sprint data.

imaging, and pay-per-view television. The main problem for the Baby Bells is to transform the monopoly mind-set they have had for all these years. The Bell companies blame the regulators for not letting them build up their investment resources, and the customers blame the operators for their unresponsive attitude to their needs. So far, regulators, judges, and lawmakers have wielded enormous power over the services market because of the monopoly enjoyed in local service.

Regulatory and legal policies in the United States have driven the Bell companies to seek investments abroad. For example, Bell Atlantic and Ameritech own 75 percent of New Zealand's phone company. Nynex, in collaboration with a Thai partner, will build a new network in Thailand with a contract to operate it for 25 years. Bell South was able to purchase a 10 percent interest in Telefonos de Mexico and already has made a paper profit of $1.5 billion. Perhaps the most promising venture is a new signaling network called Signaling System 7 that permits a whole new set of services such as Caller ID, caller tracing, and call blocking. These companies are now developing a technique for compressing full motion video for transmission over telephone lines (National Telecommunications and Information Administration Report, 1991).

In the process of deregulation of telecommunications services, most countries are reexamining the role of the regulators, because competition has taken over the work of regulation and services are being provided in a setting of freedom of entry. This more or less guarantees the efficiency of the service provider if competition really operates. The pattern varies form one country to another, as we have observed in this volume, but the goal of universal service remains in all of them. The patterns vary only in the provision of value-added services and the convergence of telecommunications with computers and broadcasting. According to Nambu (1991), it is the Japanese bureaucracy that differentiates the regulatory system in Japan from those of other countries. He characterizes the bureaucracy in Japan as a solid, integrated one that does not make use of expert systems but relies on competition among the different ministries as legitimizing policy options. In the case of telecommunications, the Ministry of Posts and Telecommunications (MPT) competes with the Ministry of International Trade and Industry (MITI) and the Ministry of Finance so that regulatory policy is guided by the bureaucrats, who become the most influential players in the game. In the case of the deregulation of NTT, MPT was in conflict with MITI but succeeded in having its own way in breaking up the monopoly of NTT. In order to provide better services to the customers, NTT spun off its Data Applications Design Center to form a new subsidiary, called NTT Data Corporation. This enabled NTT to realize better growth in its operations, which shows that competition with the subsidiary helped NTT to obtain a better allocation of resources. Even in the domestic long-distance market, competition with the New Common Carriers (NCC) has been established but leaves a large share of the market with NTT. In cities like Osaka and Tokyo, the NCCs have been able to capture lucrative markets, but in other parts of the country, they have not, simply because the price differential is 20 percent despite the protection that the NCCs receive from the government. When we examine the case of Japan, we find that its regulatory results have been closer to those in Europe than to the United States. The shift from the old monopoly paradigm has not brought the same effects in rebalancing tariffs and providing new services under widespread competition in Europe and Japan as it has in the United States.

In Canada, access to affordable basic telecommunications services is recognized as fundamental in all policy decisions. Because the advanced services of one decade become the basic services of the next, the concept of universal service is applicable to both. Canadians demand access to affordable basic telecommunications services, and policy is designed to best ensure not only the universality of basic services but also one that will stimulate the efficient development and diffusion of advanced services (Canadian Local Networks Convergence Committee, 1992). The analysis of Melody and Anderson has shown that Canada needs to retain and improve the competitiveness of its communications networks. Canada has a comparative advantage in promoting its telecommunications sector, especially in light of the North American Free Trade Agreement negotiations of 1992. The conflicting issues are of attaining maximum public benefit as against economic efficiency criteria. The trend in general has been to retain the status quo. In Canada, the historical origins and objectives of telephone and cable regulation are quite distinct. Telephone companies have been regulated as common carriers, with an emphasis on rate regulation and the prevention of abuse of monopoly power.

The OECD called on its member countries through its Committee for Information, Computer and Communications Policy to recognize the role of digital communications, under which it is no longer feasible to treat the computing, telecommunications, and broadcasting industries as separate sectors. This committee recommended that regulators should allow network sharing and that competition between multiple network providers should be encouraged. In the United States, the ban on cross-ownership of telephone and cable networks remains in effect. The NTIA has recommended that the ban be repealed and that local exchange carriers be allowed to provide video programming subject to safeguards against cross-subsidization. In Japan, local network convergence is far less of an issue than in the United States and Canada, because direct broadcasting satellites are the preferred method of broadcasting new television programming. NTT has already announced plans to build a fiberoptic network beginning in 1995 with nationwide coverage (Galbraith, 1991).

Another major player in applying the Information Revolution's technologies for economic growth has been China. Telecommunications form an integral part of their program for modernization. According to Pan Yupeng (1992), the total value added of China's Post and Telecommunications services has increased by a factor of 17 between 1978 and 1991; in 1991, it stood at 206 billion yuan. The telephone density has grown from 0.43 per 100 persons to 1.26 in 1991. The range and application of data services have also grown concomitantly with voice services. To meet the explosion in demand for data services, the switching systems are being changed from analog to digital. As shown by Lin Sun in the chapter on China, fiberoptic cable networks have grown since the 1980s, and most provinces are now able to boast fiberoptic lines. In 1991, China installed a trunk line using optic fiber with digital transmission between Najing and Wuhan.

In addition to voice and computer systems, China has boosted its telecommunications capacity with indigenously built satellites. Eleven of these supply domestic services. Television programs are now available nationwide via satellite. The Central People's Broadcast Corporation uses satellite channels; they mostly provide educational television and language programs and provide information on the economy. By 1990, about 1,600 telephone lines were using satellite retransmission, and the Bank of China communicates with its 350 branches by computers that transmit data using domestic satellites. The Long March launch system has enabled China not only to launch its own satellites but also to gain commercially from launching foreign satellites like Asiasat, Aussat, and other European satellites.

In the Association of Southeast Asian National countries the sector responses to users in the telecommunications sector have varied, depending on whether the market is strictly under a government monopoly or whether it has liberalized, giving better choices to the consumers for benefiting from the new information technologies. Malaysia has made vast strides in supplying data, cellular, and voice services under quasimonopoly systems. Likewise, Thailand is moving toward greater participation of private enterprises within and outside the country to improve its poor telecommunications facilities. Indonesia has proved beyond doubt that it could bridge the economic gap for its many islands with the Palapa system of satellites. According to Parapak (1992), the Palapa system not only has introduced unity among the different classes and regions of the country but also has led to improved educational systems and distance learning programs. A United Nations Development Program Report of 1991 indicates that the Palapa system has in general improved the quality of life through a better use of human resources, self-sufficiency in food (imports of food supplies are no longer needed), and improved family planning programs relayed over radio and television.

Despite the problems of regulators and who they regulate, the winners and losers in the information technology (IT) game are now becoming clearer. Supply driven use of IT has resulted in the composition of suppliers and users. At a meeting of the OECD's Committee for Information, Computers and Communications Policy held in Paris in October 1992, the chairman of the Committee, Richard Beard, declared that the forces of innovation have propelled the various sectors of the industry toward constant structural adjustments. Essentially, this is what the regulators have been up against under the changing demands of the users who want to benefit from the information society. Corporate users in OECD countries as well as in developing ones have made large investments in IT because of its impact on productivity and overall economic development. The growing services sectors in these countries are also keenly aware of the advantages of evolving technologies, with the result that global regulatory policies are constantly facing the challenge of change.

REFERENCES

Averch, H., & Johnson, L. (1962, December). Behavior of the firm under regulatory constraint. *American Economic Review*, *52*, 1052–69.

Beard, Richard (1992, December). Chairman OECD's IC CP Committee cited in *Telecommunications and data report*, Virginia. May/June edition, p. 15.

Business Week. (1992, October 5). Article on Regional Bell Operating Companies.

Canadian Local Networks Convergence Committee. (1992). *Report on convergence, competition and cooperation*. Ottawa: Ministry of Supply and Services, pp. 40–50.

Carpentier, M. (1991). The single European market and telecommunications in a world context. In *Single Market Communications Review* (p. 28). London: Kline.

Dizard, W. (1992). The EC-92 telecommunications sweepstakes: Who won? Who lost? What next? In Cafruny, A., & Rosenthal, G. (Eds.), *The state of the European community*. London: Longman.

The Economist. (1992, November 14). Article on The Baby Bells.

Galbraith, M. (1991, May 6). Japan thinks big on the fibre front. *Telephony*.

Jussawalla, M. (1985). Productivity and information technology. *PTC Quarterly*, *6*, 1–6.

Jussawalla, M., & Cheah, C. (1987). *The calculus of international communications: A study in the political economy of transborder data flows*. Boulder: Libraries Unlimited.

Lamberton, D. M. (1983). *The trouble with technology*. London: Pinter Press.

Melody, W. (1990). Future world markets for information technology. In

J. Mueller (Ed.), *IT: Impacts, policies and future perspectives: Proceedings of TIDE 2000 conference in Berlin*. Berlin: Springer-Verlag.

Mosco, V. (Ed.). (1984). *Policy research in telecommunications: Proceedings from the eleventh annual conference*. Norwood: Ablex.

Nambu, T. (1991). Regulatory issues. In Chamoux, J. P. (Ed.). *Deregulating regulators*. Amsterdam: IOS Press.

National Telecommunications and Information Industry. (1991). *NTIA report: Telecommunications in the age of information 1991*. Washington, D.C.: Department of Commerce.

Noam, E. M. (Ed.). (1983). *Telecommunications regulation today and tomorrow*. New York: Harcourt Brace Jovanovich.

National Telecommunications and Information Administration. (1991). *Telecommunications in the age of information*. Washington, D.C.: Department of Commerce.

Pan, Y. (1992, November/December). IT and telecom keys to promoting China's economic reform, TDR, pp. 25–27, Virginia, U.S.A.

Parapak, J. L. (1992). Role of IT and telecommunications in Indonesian development in the 1990s. In *East Asian economic development proceedings of TIDE 2000 conference*. Seoul. Amsterdam: TIDE 2000 Secretariat.

Peltzman, S. (1976, August). Toward a more general theory of regulation. *Journal of Law and Economics, 19*, 211–40.

____. (1981). Current developments in the economics of regulation. In Fromm, G. (Ed.), *Studies in public regulation* (pp. 371–84). Cambridge: MIT Press.

Snow, M., & Jussawalla, M. (1981). *Telecommunications economics and international regulatory policy*. Westport: Greenwood.

Stigler, G. (1971, Spring). The theory of economic regulation. *Bell Journal of Economics and Management Science, 2*, 3–21.

Selected Bibliography

Baughcum, Alan. "Deregulation, Divestiture, and Competition in U.S. Telecommunications: Lessons for Other Countries." In *Marketplace for Telecommunications: Regulation and Deregulation in Industrialized Democracies*, edited by Marcellus S. Snow, pp. 69–105. New York: Longman, 1986.

Bolter, Walter G. "The Continuing Role of Federal Regulation in the Transition to Competition in Communications." In *Issues in Public Utility Regulation*, edited by Harry M. Trebing, pp. 401–17. East Lansing: Michigan State University Press, 1979.

Borchardt, Kurt. *Structure and Performance of the U.S. Communications Industry: Government Regulation and Company Planning*. Boston: Harvard Business School, 1970.

Branscomb, Anne W., and George J. Lissandrello. "A View of Telecommunications Policies in the United States." In *Pacific Telecommunications Conference Proceedings 1980*, edited by Dan J. Wedemeyer, pp. 2C-1–14. Honolulu: Pacific Telecommunications Council, 1980.

Brock, Gerald W. *The Telecommunications Industry*. Cambridge, MA: Harvard University Press, 1981.

Bruce, Robert R. "Implementing United States International Telecommunications Policy: Observations on Recent Developments and the Need for New Institutional Arrangements." In *Proceedings from the Tenth Annual Telecommunications Research Conference*, edited by Oscar H. GandyJr., Paul Espinoza, and Janusz A. Ordover, pp. 267–83. Norwood: Ablex, 1983.

Crandall, Robert W. "Deregulation: The U.S. Experience." *Zeitschrift fur die gesamte Staatswinssenchaft* 139 (November 1983): 419–34.

DeRosa, Francis J. "Limitations for the Entry of New Carriers." In *The Washington Round: World Telecommunications Forum, Washington, DC, 18–19 April 1985*, pp. 61–76. Geneva: International Telecommunication Union, 1985.

Eward, Ronald S. *The Competition for Markets in International Telecommunications*. Dedham: Artech House, 1984.

____. *The Deregulation of International Telecommunications*. Dedham: Artech House, 1985.

Frieden, Robert M. "The International Application of the Second Computer Inquiry." In *Regulation of Transnational Communications*, edited by Leslie J. Anderson, pp. 189–218. New York: Clark Boardman, 1984.

Geller, Henry. "The New Telecommunications Act as a Regulatory Framework." In *Telecommunications Regulation Today and Tomorrow*, edited by Eli M. Noam, pp. 205–55. New York: Harcourt Brace Jovanovich, 1983.

Goldberg, Henry. "International Telecommunication Regulation." In *Communications for Tomorrow: Policy Perspectives for the 1980s*, edited by Glen O. Robinson, pp. 157–87. New York: Praeger, 1978.

Irwin, Manley R. *The Telecommunications Industry: Integration vs. Competition*. New York: Praeger, 1971.

____. "U.S. Telecommunications Policy: Beyond Regulation." In *Kommunikation ohne Monopole: Ueber Legitimation und Grenzen des Fernmeldemonopols*, edited by Ernst-Joachim Mestmacker, pp. 51–72. Baden-Baden: Nomos, 1980.

Johnson, Leland L. "Technological Advance and Market Structure in Domestic Telecommunications." *American Economic Review* 60 (May 1970): 204–08.

Johnson, Nicholas, and J. J. Dystel. "A Day in the Life: The Federal Communications Commission." *Yale Law Journal* 82 (July 1973): 1575–1634.

Langdale, John. "Competition in the United States' Long-Distance Telecommunications Industry." *Regional Studies* 17 (December 1983): 393–409.

Lewin, Leonard (ed.). *Telecommunications in the U.S.: Trends and Policies*. Dedham: Artech House, 1981.

Meyer, John R., et al. *The Economies of Competition in the Telecommunications Industry*. Cambridge, MA: Oelgeschlager, Gunn & Hain, 1980.

Militzer, Kenneth, and Martin Wolf. "Deregulation in Telecommunications." *Business Economics* 20 (July 1985): 27–33.

Mosco, Vincent. "Who Makes U.S. Government Policy in World Communication?" *Journal of Communication* 29 (Winter 1979): 158–64.

Noll, Roger G. "'Let Them Make Toll Calls': A State Legislator's Lament." *American Economic Review* 75 (May 1985): 52–56.

Polsby, Daniel, and Kim Degnan. "Institutions for Communications Policymaking: A Review." In *Communications for Tomorrow: Policy Perspectives for the 1980s*, edited by Glen O. Robinson, pp. 501–14. New York: Praeger, 1978.

Robinson, Glen O. "The Federal Communications Commission." In *Communications for Tomorrow: Policy Perspectives for the 1980s*, edited by Glen O. Robinson, pp. 353–400. New York: Praeger, 1978.

Sidel, M. Kent, and Vincent Mosco. "U.S. Communications Policy Making: The Result of Executive Branch Reorganization." *Telecommunications Policy* 2 (September 1978): 211–17.

Stanley, Kenneth B. "International Telecommunications Industry: Interdependence of Market Structure and Performance Under Regulation." *Land Economics* 49 (November 1973): 391–402.

Trauth, Eileen M., Denise M. Trauth, and John L. Huffman. "Impact of Deregulation on Marketplace Diversity in the U.S.A." *Telecommunications Policy* 7 (June 1983): 111–20.

Trebbing, Harry M. "A Critique of Structure Regulation in Common Carrier Telecommunications." In *Telecommunications Regulation Today and Tomorrow*, edited by Eli M. Noam, pp. 125–76. New York: Harcourt Brace Jovanovich, 1983.

____. "Public Utility Regulation: A Case Study in the Debate over Effectiveness of Economic Regulation." *Journal of Economic Issues* 18 (March 1984): 223–50.

Webbink, Douglas W. "The Recent Deregulatory Movement at the FCC." In *Telecommunications in the U.S.: Trends and Policies*, edited by Leonard Lewin, pp. 75–83. Dedham: Artech House, 1981.

Wiley, Richard E. "Competition and Deregulation in Telecommunications: The American Experience." In *Kommunikation ohne Monopole: Ueber Legitimation und Grenzen des Fernmeldemonopols*, edited by Ernst-Joachim Mestmacker, pp. 31–50. Baden-Baden: Nomos, 1980.

Index

About the Editor and Contributors

Meheroo Jussawalla is a Senior Research Fellow and Economist in the Program on Communications and Journalism at the East-West Center in Honolulu. She has done pioneering work in the economics of information and telecommunications as related to development in the Asia-Pacific region. She has published 12 books and many articles in accredited journals. Her latest books are *The Economics of Intellectual Property Rights in the World Without Frontiers* (Greenwood, 1992), and *United States-Japan Trade in Telecommunications: Conflict and Compromise* (Greenwood, 1993). She is on the Advisory Committee of *Transnational Data and Communications Report* and on the Editorial Borad of *Information Economics and policy*. She is an elected member of the Board of Trustees of the International Institute of Communications (London) and the Board of Trustees of the Pacific Telecommunications Council (Honolulu).

Raymond U. Akwule is a Professor in the Department of Communication at George Mason University, Fairfax, Virginia. He has worked in Africa and published case studies in African telecommunications.

Peter S. Anderson is a Professor of Communications at Simon Frazer University in Vancouver, B.C., Canada.

Raimundo Beca is a Telecommunications Adviser to the Companía de Teléfonos de Chile in Santiago, Chile. He recently was with the UN Economic Commission for Latin America as a Regional Expert.

Jean-Pierre Chamoux worked with the Regulatory Commission for Telecommunications in France. His expertise is in the regulatory and legal aspects of telecommunications in Europe. He is the Editor of *La Communicateur*, a quarterly journal published in Paris.

Henry Geller is Director of the Washington Center for Public Policy Research, Duke University, Institute of Policy Sciences and Public Affairs. He was formerly Assistant Secretary of Commerce for Communications and Information — National Telecommunications and Information Administration until 1981. He was General Counsel to the Federal Communications Commission and became Special Assistant to the Chairman of the Federal Communications Commission.

Klaus Grewlich is the Director General for International Affairs Telekom, Deutsche Bundespost, Germany. He publishes widely on issues of international communications.

Heather E. Hudson is a Professor at the McLaren School of Business at the University of San Francisco. She is well-known for her research and publications dealing with telecommunications for remote and rural areas. She is also an expert in satellite communications.

D. McL. Lamberton is a pioneer in information economics and was a co-founder of the Center for International Research in Communication and Information Technologies in Australia. He is now a Visiting Fellow in the Urban Research Program at the Research School of Social Sciences at the Australian National University in Canberra.

William Melody is a Founder and Director of the Center for International Research in Communication and Information Technologies in Melbourne, Australia.

Adrian Norman is an internationally recognized Telecommunications Consultant. His company is now Adrian Norman and Associates in Oxford, England. Prior to this, he worked with Arthur D. Little and the SEMA Group, London.

Hajime Oniki is an Economist specializing in telecommunications. He is the Director of the Institue for Social and Economic Research at Osaka University, Japan.

Lin Sun is a Telecommunications Consultant with COMSTREAM in San Diego. He is based in Beijing, China. He has recently received his Ph.D. from the State University of Michigan.